THE SPEED OF LIGHT

DOCUMENTATION
Name:
Ship:
Dates of Voyage:
Stateroom:

THE SPEED OF LIGHT

LOUISE RIOFRIO

Dedicated to

Mother and Father,

who brought me into the light.

First Edition May 2014

Second Edition March 2016

Third Edition June 2021

Contents

Acknowledgements

Great thanks are due to colleagues at Johnson Space Center, especially Dr. Bonnie Cooper and the late Dr. David McKay. The late Dr. Marni Dee Sheppeard was a tireless supporter. Marianne Dyson, Sherry Crowson, and Denise Little provided commentary on the first chapter. The photographs taken from Space are courtesy of NASA.

I wish to thank the backers of the crowdfunding project that helped get this book published. Special thanks to Carl Brannen, Christopher Fedde, Talia Peschka, Ed Minchau, Fred Chavez, Andrew Stevens, Scott Early, Markus Jacquemain, Ryan Pederson, Susan and John Husisian, George Stelzenmuller, Julian Calliandro, Tristan Bettencourt, Andrea and Arthur Hoekstra, and Nic Gihl.

Introductory Essay

Professor Avi Loeb, PhD.
Frank B. Baird Professor of Science
Chair of Astronomy Department, Harvard University

Recently, my family and I visited the Mayan city of Chichen Itza in Mexico, one of the New 7 Wonders of the World. As our tour guide marveled on the spectacular astronomical measurements of the Mayan culture, I began to wonder why our modern scientific understanding of Astronomy did not originate in South America. After reading some historical background, I realized that unfortunately the Mayans had used their exquisite astronomical data within a mythological culture of astrology that rested upon false but mathematically sophisticated theories about the Universe.

They collected unprecedented amounts of precise astronomical data on the Sun, the moon, the planets and the stars, but failed to come up with the breakthrough ideas of Nicolaus Copernicus, Galileo Galilei, Johannes Kepler and Isaac Newton. The sober realization that scientific advances can be trapped by cultural and societal forces inspired me to wonder: have we learned the necessary lessons to prevent our current scientific culture from resembling Mayan Astronomy?

Summit of pyramid at Chichen Itza, Yucatan

Let me sharpen the question further: is data collection by itself a guarantee for good science? This issue has important implications regarding the effectiveness of allocating resources by private and federal funding agencies for the advancement of science.

Based on the historical success of empirically-based deduction in modern science, one would naively expect that a culture dedicated to the collection of the highest quality data would inevitably arrive at the proper scientific interpretation of this data and unravel the appropriate theory for making predictions that are testable by future data. With that as our model, all we would need to do in order to cultivate a productive scientific endeavor would be to fund state-of-the-art experimental or observational programs and scientific advances will simply follow. This popular strategy in funding science currently guides the allocation of most of the Astronomy Division funds at the US National Science Foundation (NSF) to major facilities and large scale surveys. The focus is clearly on large team efforts to collect better data within the mainstream paradigms of Astronomy, under the assumption that good science will follow.

The Mayan example leads to the realization, however, that a successful scientific culture requires a separate guiding principle, namely the desire to compare multiple interpretations of existing data and multiple motivations for collecting new data. This additional principle fosters scientific debate among competing theories, and encourages free thought outside the mainstream. Without this added principle, a culture could be scientifically inefficient at interpreting existing data and misguided in its motivation for collecting new data.

The Mayan culture collected exquisite astronomical data for over a millennium with the false motivation that such data would help predict its societal future. This notion of astrology prevented the advanced Mayan civilization from developing a correct scientific interpretation of the data and led to primitive rituals such as the sacrifice of humans and acts of war in relation to the motions of the Sun and the planets, particularly Venus, on the sky.

Given the strong sociological trends in the current funding climate of team efforts, how could we reduce the risk of replicating the indoctrinated Mayan astronomy? The answer is simple: by funding multiple approaches to analyzing data and multiple motivations to collecting new data. After all, the standard model of cosmology is merely a precise account of our ignorance: we do not understand the nature of inflation, the nature of dark matter or dark energy.

Our model has difficulties accounting for what we see in galaxies (attributed often to complicated "baryonic physics"), while at the same time not being able to see directly what we can easily calculate (dark matter and dark energy). The only way to figure out if we are on the wrong path is to encourage competing interpretations of the known data. When science funding is tight, a special effort should be made to advance not only the mainstream dogma but also its alternatives.

To avoid stagnation and nurture a vibrant scientific culture, a research frontier should always maintain at least two ways of interpreting data so that new experiments will aim to select the correct one. A healthy dialogue between different points of view should be fostered through conferences that discuss conceptual issues and not just experimental results and phenomenology, as often is the case currently. These are all simple, off-the-shelf remedies to avoid the scientific misfortune of the otherwise admirable Mayan civilization.

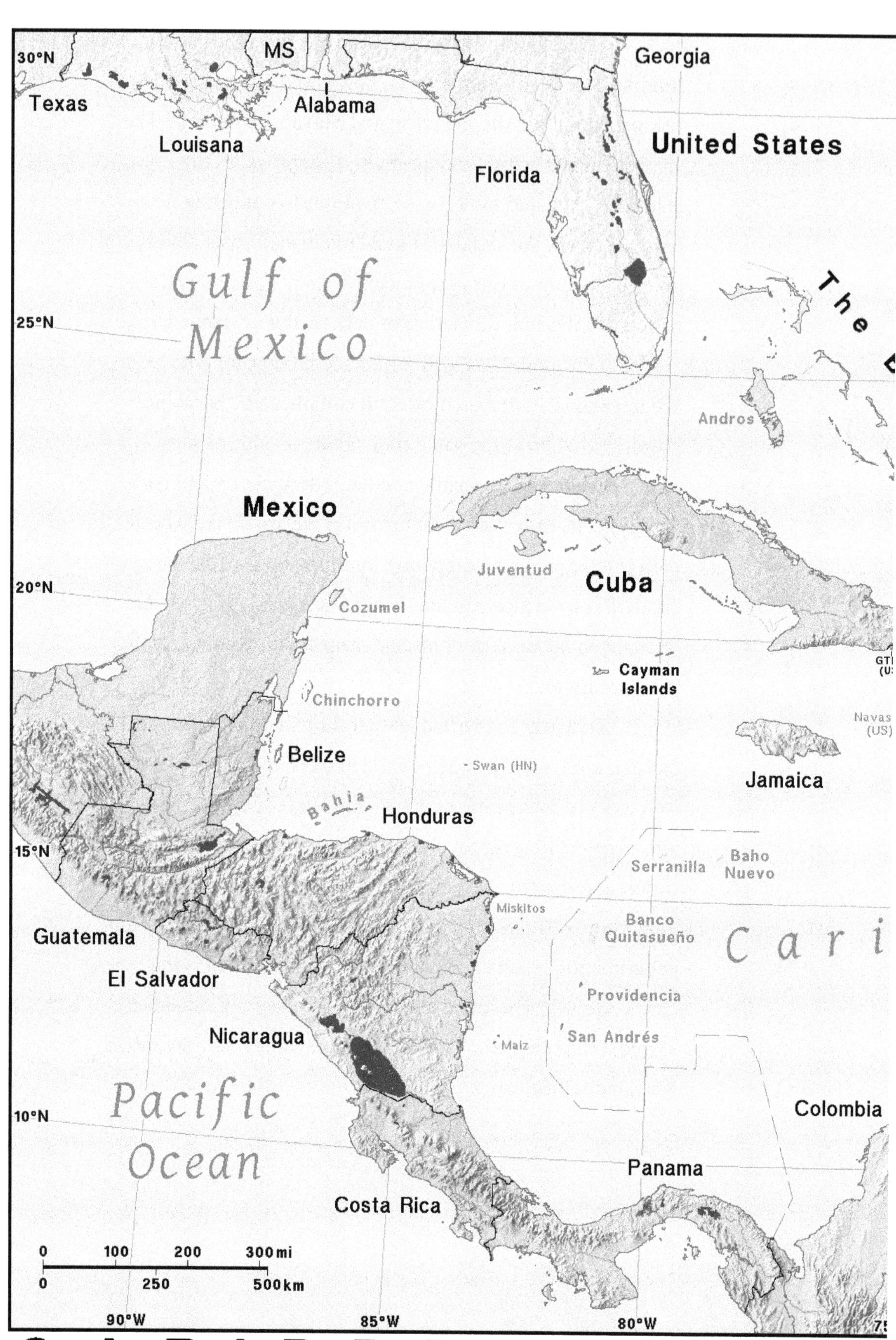

C A R I B B E A N S E A

2015
Sargasso
Sea
Atlantic
Ocean
Bahamas
Turks and
Caicos Islands
Inagua
Dominican
Republic
Haiti
Hispaniola
Virgin
Islands
Anguilla
St
Martin
St Barthélemy
Barbuda
Antigua
Mona
(PR)
Puerto
Rico
St Kitts
and Nevis
Montserrat
Guadeloupe
Dominica
Martinique
St Lucia
bbean Sea
St Vincent
and the
Grenadines
Barbados
Netherlands
Curaçao
Aruba
Bonaire
Aves
Orchilla
Blanquilla
Grenada
Roques
Margarita
Tobago
Tortuga
Trinidad
Venezuela
Guy.
70°W
65°W
60°W

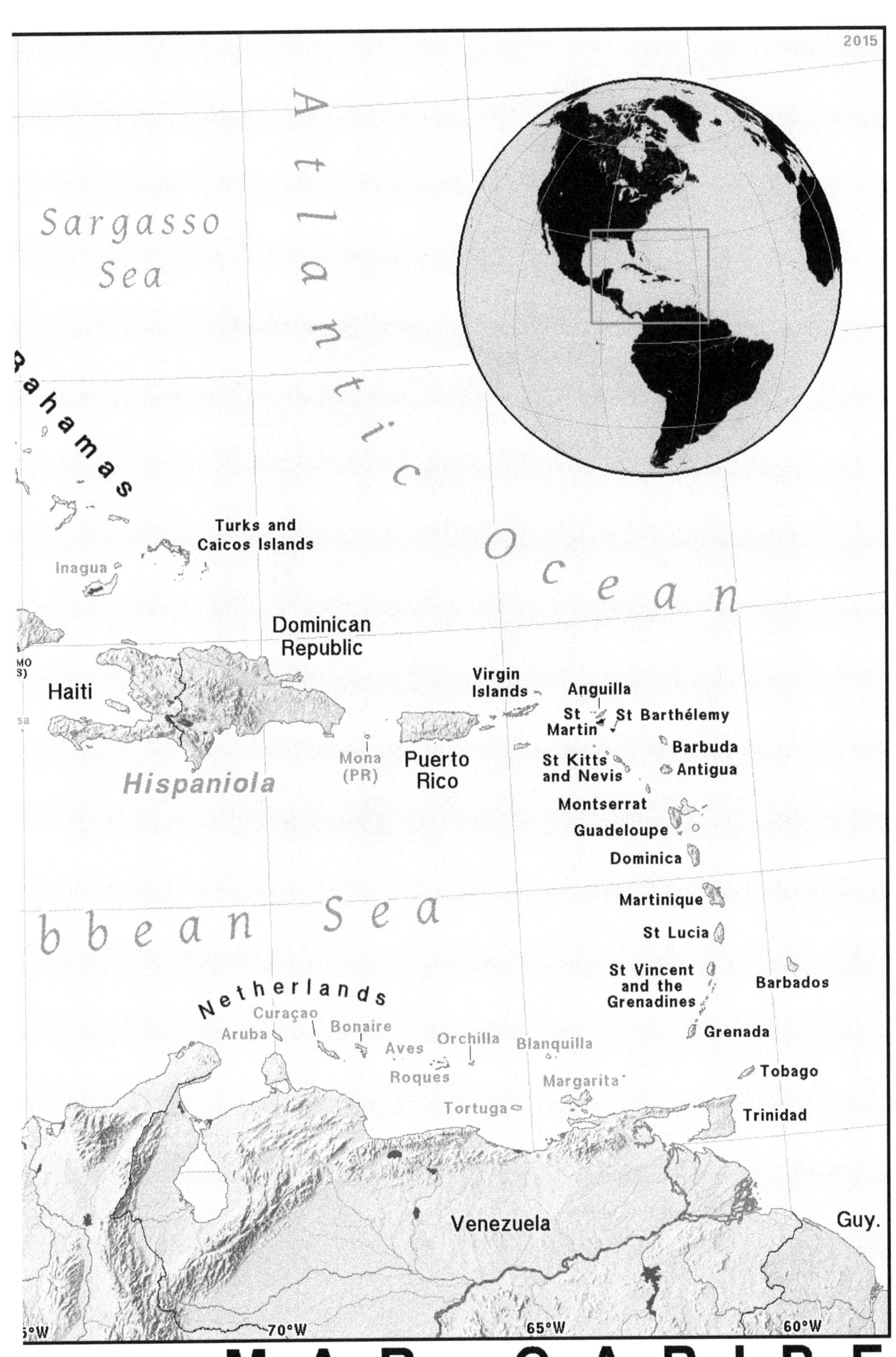

• MAR CARIBE

Observatory at Chichen Itza

1
A Babe in the Universe

A child's eyes open, and light floods in.

Our lives begin in darkness, the bubble of the womb. Within its confines a baby experiences warmth and pain without the light of understanding. At birth she emerges into a far larger Universe. With her first gasp a baby breathes an atmosphere transparent to light. Like a book opening, the Universe is revealed. Through light her eyes will see much of this world. Though it may seem cold and dark, we live in a Universe of light.

Each human birth is the culmination of many events. A baby has grown from a tiny cell barely a millionth of a meter across. Before her parents met, a parade of ancestors walked the Earth. Her human species resulted from evolution of life over billions of years. Before life could take hold on Earth, the Sun and Solar System formed in Space. A Milky Way galaxy formed with hundreds of millions of stars. The galaxy is part of an enormous Universe that also grew from a tiny point. Like the tip of a cone, many events converge to create a single human life.

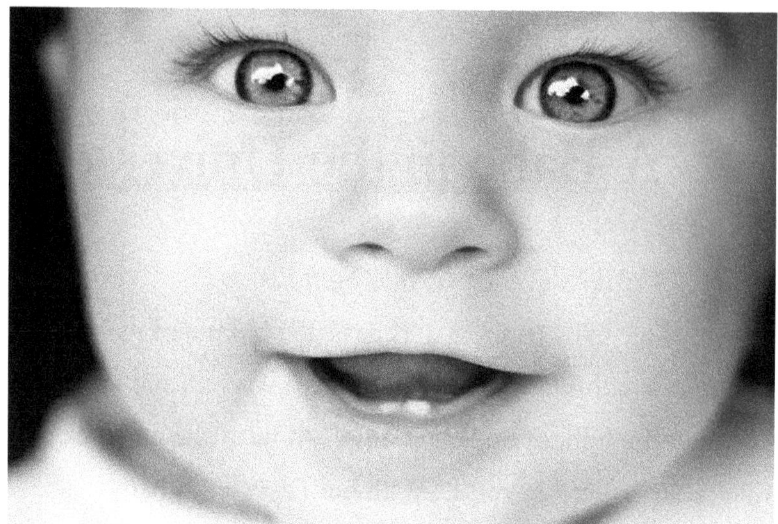

Astronomers celebrate "First Light" when a new telescope is first opened to the Universe. A baby experiences First Light when her eyes open. Images, sounds and language rapidly accumulate in her growing brain. Watching a child grow is witnessing the process of discovery.

Much of a baby's first months are spent staring, especially at faces. Infants are quite nearsighted, and can focus no more than 30 centimeters away. Almost immediately she can recognise faces that come into view. Humans are so well programmed to find faces that we imagine seeing them in the surface of the Moon or Mars. Discovery causes a baby to smile--an expression so universal that it need not be taught.

During a baby's first few months she learns causes and effects. She finds that crying can bring a response from others. She becomes conscious of her hands and feet, and begins to learn their uses. She learns about gravity by falling down many times. An instinct to explore drives her to crawl and eventually to walk.

From this curiosity comes a lifetime desire to learn. It would be impossible to learn all of knowledge from scratch, so humans build upon the lessons of others. Humans often tend to do things the way they have been done before, and repeat answers they have heard before. In this manner knowledge is passed from one generation to another.

Language is another important skill, for the world seems less mysterious when things have names. Once a thing is named, she can attach other words to further describe it. She may someday learn mathematics to describe the world more precisely. These languages of words or numbers are tools for understanding.

She can be heard talking quietly in a language that only a baby understands. Even after learning from others, she remains capable of finding her own descriptions of things. The ability to find her own answers is a key to expanding human knowledge. Time and again, we will see that advances begin from the light of a single individual.

Around her second birthday a child has earned concepts of Space, can associate the kitchen with eating and her bedroom with sleeping. She still lives mostly in the present, seeming to want everything right now! By her third birthday, she has learned about before and after, the beginnings of her understanding of Time. Much later humans learn to connect Space and Time, coordinating their actions to catch a ball or arrive at a place on time. We will see that Space and Time are linked by the Speed of Light.

Pictures in a book are an early source of interest. Her eyes unconsciously connect the two dimensions of a picture with the three dimensions of her world—the picture of an apple is an apple in her imagination. Pictures in books illuminate a new and magical world for her eyes to see. The writing beneath the pictures soon draws interest, too. Through reading a child can access the sum of human experience. Opening a book brings light to the world.

A child's mind is full of questions. The curious child will repeatedly ask her parents or teachers, "why?" An answer will just get another "why?" as the child seeks a deeper meaning. A new object or animal elicits a curious "What's that?" A mind that continues learning about the Universe also continues to grow.

When faced with something new, humans often find temporary answers. At some time a child may be told that babies are delivered by a stork. Later she learns the truth is far more exciting. Humans have often adopted temporary explanations, such as believing that Earth is centre of the Universe. Growth requires the ability to abandon a theory when a better one comes along. The light of knowledge advances when theories are improved.

Much of human history was spent huddling in shelter at night with darkness all around. The only light would have been the Moon, stars and an occasional firefly. Creatures lurking in the night seemed mysterious and dangerous. Humans are instinctively drawn to light and knowledge. The rising Sun or the fires of home are sources of comfort and warmth. Scientists have spent centuries studying, arguing and wondering about light.

The process of discovery follows a pattern learned by babes. We are presented with a world that at first seems confusing. Gradually, we extend our senses and minds to understand the Universe. We learn to see patterns, connections, causes and effects. Often we learn from lessons of others before us. Sometimes we adopt a temporary explanation until a better one comes along.

When an individual comes up with an idea, the lesson may be passed on to others and all humanity advances. A baby's influence radiates outward like starlight. Eventually her life affects a whole community. As she grows into adulthood, a human has the potential to affect events across the planet and possibly beyond.

Humans share many of the same questions, starting with who they are and what is their place in the world. How big is the world and what is its shape? How did the world come to exist? How were the Sun, Moon and other sky objects created? How did the Universe begin? These questions are so universal that they have been asked across history.

A child's curiosity leads to a history of exploration. We will start with the discovery that Earth is curved and not flat. The question whether Earth was centre of the Universe is an important lesson for today's science. From there we move outward to discoveries about the nature of Space and Time. The phenomenon of light is an intimate part of this Universe. We will see old questions about light itself, such as its speed or whether it forms waves or particles. The phenomenon of light is key to discoveries from Relativity to the tiny quantum world. We will see that our Universe began with a burst of light.

I had the privilege of predicting and measuring change in the Speed of Light, a discovery that changes everything. From studies of the Moon at NASA, evidence was found that light slowed exactly as predicted. Some of the data was recently published in a refereed scientific journal, and will be described in Chapter 8. Light is key to understanding our Universe.

Our exploration of the Universe begins at First Light. The world we experience is the result of many events coming together. Just as a child grows, human knowledge also grows. Discoveries about this Universe come with the Speed of Light.

NEXT: The Light of Exploration

2
The Light of Exploration

We can easily forgive a child who is afraid of the dark; the real tragedy of life is when men are afraid of the light.-- Plato

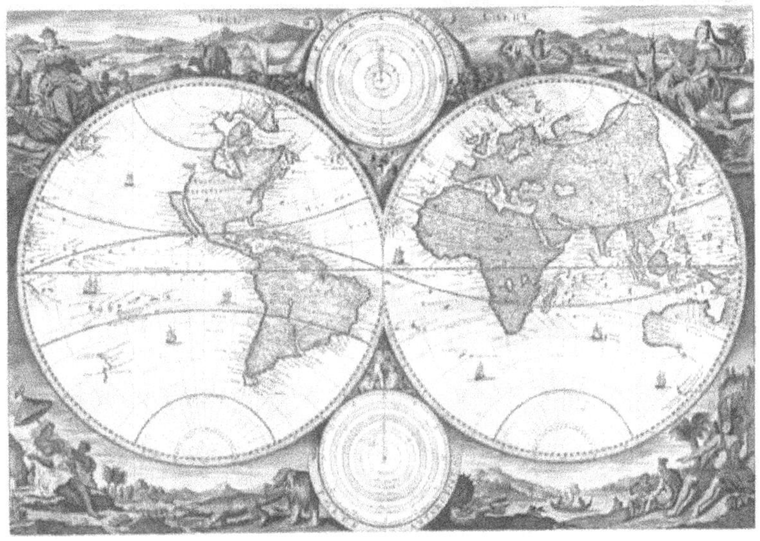

Map of the world by Jan Stoopendaal, 1730

Life begins in the sphere of the womb, but once upon a time, humans may have thought their world was flat. That was an understandable assumption when their "world" was very small. For most of human history, few people ventured further than a few kilometres from their place of birth. For planning a home, farm, or even a town we need not take into account Earth's curvature.

Even if one lives on a plain or by the seashore, the horizon appears to vanish into the distance like an infinite flat plane. For most of the humans who have lived, it was unnecessary to think of Earth's shape. Scientists have at times thought that the entire Universe is flat, like the Earth.

She began life as a curious child, then grew into a beautiful and intelligent woman. Despite the responsibilities of children and adulthood, she retained her curiosity about the world. Many days she spent staring at the sea, wondering what lay beyond. Having seen babies grow into adulthood, she wondered whether the world had a size and shape. Though the woman's name is lost to history, she made one of the greatest discoveries of all.

The theory that Earth is round may have been first made by a woman waiting for a ship to return. Peering anxiously at the horizon, she would have seen a ship's sails appear before the hull. While waiting for days, she may have contemplated the curve of her breasts, seeing how a point on her skin could disappear around their curve. If she were pregnant, she would have seen her belly grow from flat to round. Comparing her observations, she might have suspected that she lived on the surface of an immense sphere.

From the point-of-view of a sailor on the ship, the land that he has set out from sank below the horizon. The sight of one's home disappearing into the sea must have been quite alarming! The first sailors to witness this effect must have turned and hurried home. Upon the sailor's return, the woman would claim that it was the ship that appeared to sink. They had both disappeared *Relative* to each other's frame of reference. Discovery of Earth's curvature began a quest by men and women that would someday lead to a Theory of Relativity.

Since the discovery of Earth's shape, many women have waited for their ship to come in. They have waited for exploration to bring food to the table or a better life for their children. They have waited for a time when their horizons would be greater—when a woman's ideas were heard. That ship may soon arrive.

The Mediterranean Sea

Sailing the Mediterranean, you are quickly out of sight of land and in open sea. To ancient mariners it seemed endless, the "Middle of the Earth". The Phoenicians, a seagoing people, are thought to be the first to have navigated beyond the horizon. Movements of the Sun and Moon determine East and West. The star Polaris, around which other stars seem to rotate overnight, points the way North. Astronomy was also used to predict solstices as an aid to farming. Knowledge of the stars led to a better life on Earth. More about navigation is in my book DISCOVERY: ALASKA TO HAWAII.

2015
Atlantic Ocean
Bay of Biscay
Germany
Switzer.
SM
Monaco
France
Corsica
VA
Portugal
Spain
Minorca
Majorca
Ibiza
Sardinia
Tyrrhenian Sea
Mediterranean Sea
Strait of Gibraltar
Gibraltar
Algeria
Morocco
Tunisia
MR
Mali
5°W
0°
5°E
10°E

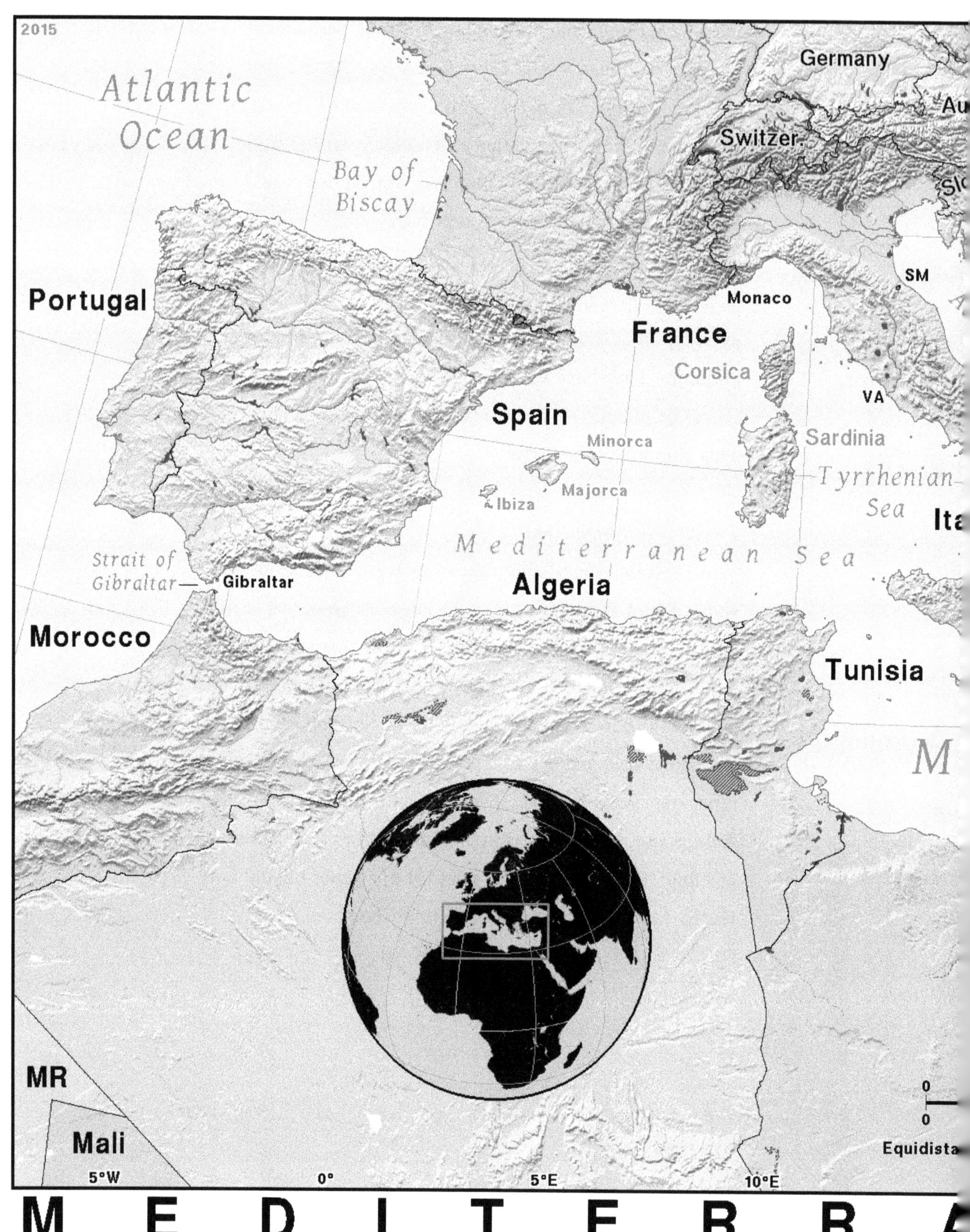

M E D I T E R R A

Ukraine
Moldova
Russia
45°N
tria
Hungary
enia
Crimea
Croatia
Romania
Black Sea
Georgia
Bosnia
and
Herz.
Serbia
Bulgaria
Turkey
ME
40°N
iatic Sea
Bosporus
Albania
Mace.
Ionian
Sea
Syria
35°N
Sicily
Greece
Cyprus
Malta
Crete
Lebanon
editerranean Sea
Israel
WB
Gaza
Libya
Jordan
Egypt
30°N
Gulf of Sidra
Saudi
Arabia
Red
Sea
200
400 mi
25°N
250
500 km
Conic Projection: SP 30°N/45°N
°E
20°E
25°E
30°E
35°E

N E A N S E A

Statue of Pythagoras on island of Samos

In school we hear about Pythagoras via a theorem about triangles, but he was also interested in spheres. Pythagoras was born around 580 B.C. on the island of Samos. Today on an excursion to Samos you can see his statue in a village called Pythagoreio, with a view of the Turkish coast across the sea. His ideas are the basis of modern physics, including a spherical Earth.

Pythagoras travelled the world in search of knowledge, from Greece to Egypt and possibly as far as India. The so-called Theorem of Pythagoras was known in Babylon and India centuries before Pythagoras lived. He may have learned it in his travels and spread it to the West. Pythagoras encouraged men and women to have diverse interests, encouraging them to explore both music and astronomy.

To please his musician's ear, Pythagoras sought a "cosmic harmony." As a musician, he is credited with the idea that "music of the spheres" described the planets. Reasoning that it was the most harmonious shape, he theorised that Earth was spherical. Pythagoras and his followers even thought that Earth revolves around a central fire, the Sun.

Pythagoras had a wife called Theano. She was an accomplished scholar in her own right, publishing works on a variety of subjects. Her work *On Piety* mentions that Pythagoras thought that all things come from numbers, meaning that everything in the world could be described by equations. Pythagorean ideas began a quest that would last thousands of years, to find equations describing the Universe.

The Agora in Athens with Acropolis in background

Athens, Greece

On an excursion to Athens today, you are in the footsteps of the Greek philosophers. The Acropolis dominates the city as it has for thousands of years. The Parthenon and its surrounding museums remind us that Greece was once the centre of Western thought. The ancient Agora was once the city's centre of business, shopping, seeing and being seen.

A century after Pythagoras, the great thinker Socrates taught his followers to question what they were taught. Socrates, who was born around 470 B.C., lived his entire life in Athens. He had no school, but haunted the Agora constantly questioning the beliefs of others. In his wake he gained a following of young people who enjoyed seeing him challenge the privileged classes.

If you are not in Athens, there are other ways to follow in Socrates' footsteps. He taught us to question everything, even what is "commonly known". If today we are told that the universe is filled with epicycles, cosmic strings or "dark" energies; we should question what we hear. Accused of corrupting the young, Socrates was imprisoned and forced to drink a poison hemlock.

Socrates left no writings of his own—all we know of him is from the works of his followers. Plato wrote that Socrates' prison was within walking distance of the Acropolis. Nearby in Philopappou Hill, also known as the Hill of Muses, are some cell-like caves carved from rock many centuries ago. A sign marks these caves as "Prison of Socrates," though we are not certain if this is where he was locked up.

Plato's Academy envisioned by Raphael.

Plato, one of the young followers of Socrates, founded his Academy in Athens just outside the city walls. You may stroll its paths in Plato's Academy Park at Monasteriou 140. Around the site are many stones among which Plato or his students could have sat. In the tree-shaded grounds, you can easily imagine Plato walking alongside discussing philosophy or science. One of Plato's sayings has great import today:

We can easily forgive a child who is afraid of the dark; the real tragedy of life is when men are afraid of the light.

Plato's Academy admitted both men and women. Two women are known to have attended, Axiothea and Lasthenia. Plato thought that the world could be deduced from simple principles. In search of those principles, he tried to relate the planets to geometric shapes. Plato studied the writings about Pythagoras, and taught the spherical Earth to his own students.

Plato's student Aristotle founded his own school, called the Lyceum. The ruins of the Lyceum were uncovered in 1996. You can see them near the intersection of Rigillis and Vasilissis Streets, close to the National Gardens and the Athens War Museum is also next door. The remains of the Gymnasium, where students trained for combat, are easily visible.

Aristotle compiled more evidence of a spherical Earth. The southern constellations appear to rise higher as one travels farther south. The shadow of Earth upon the Moon during a lunar eclipse is circular. Aristotle was fascinated by symmetry and repetition. Teacher and disciple relied on different methods--while Plato preferred to deduce the world from first principles, Aristotle looked at physical evidence. As the teachings of Plato and Aristotle have spread through time, cosmologies based upon first principles have sometimes conflicted with those assembled from measurements.

Aristotle believed that the heavens were filled with a substance called *quintessence*, which was invisible and filled all of Space. He would not be the last to assume that the Universe is filled with an invisible substance. Even today scientists are tempted to explain the universe with quintessence. Aristotle also assumed a cosmology where Earth was the centre of all things. This flawed view of the Universe would last nearly 2000 years.

One of Aristotle's students was the restless son of King Phillip of Macedonia, who later would be known as Alexander the Great. His conquests would spread Greek learning throughout the known world. After his conquest of Egypt, he founded the city of Alexandria. The centre of learning sailed east to a library in Egypt.

The Library of Alexandria rises again

Alexandria, Egypt

In 2002 a modern Library of Alexandria was opened on the shore of the Mediterranean. It is an excellent place to visit and learn about the original. The Library of Alexandria was founded around 295 B.C. by Demetrius of Phalerum, another student of Aristotle. He convinced the city's rulers that Alexandria could succeed Athens as a centre of learning.

Ships entering Alexandria's harbour saw their scrolls and maps seized and copied. Eventually the Library collected around 700,000 scrolls on everything from religion to geography. So many scrolls were collected that a second repository was opened at the temple of Serapis. More than just a collection of scrolls, it also served as a university, attracting scholars from around the Alexandrian world.

In one place were Hebrew scriptures, Buddhist texts, Egyptian astronomy, along with the complete plays of Sophocles and Euripides. Long before the internet, librarians at Alexandria had access to the amassed knowledge of the ancient world.

Aristarchus, an astronomer and mathematician from the same island of Samos as Pythagoras, lived from about 310 to 230 BC. His work *On the Sizes and Distances of the Sun and Moon* contained estimates that were strikingly good for their time. He calculated that the Moon's distance was 30 times Earth's diameter. Those distances were estimated from the Earth, fitting an Earth-centred cosmology. The enormous distances may have started Aristarchus thinking about alternatives. He wrote another book that is now lost and known only through other writers.

This second book introduced a cosmology with the Sun in the centre and Earth circling as a planet. Aristarchus also believed that the stars were extremely distant, so far that they appeared fixed in the sky. This second book caused great controversy, with another writer angrily suggesting that Aristarchus should have been put on trial! Aristarchus' cosmology was incredibly ahead its time.

By 240 BC, many educated people believed that Earth was spherical. In that year Erastothenes, Librarian of Alexandria, made a remarkable estimate of Earth's size. Making use of Pythagorean geometry, Eratosthenes combined the altitude of the Sun at different locations on

Earth with estimates of the distance to those locations. He derived a circumference of 250,000 stadia. Though the exact length of a Greek "stadium" is not known today, Erastothenes' figure was amazingly accurate.

Hipparchus, another scholar who lived from about 190 to 120 BC, calculated the distance between Earth and the Moon and also believed that Earth circled the Sun. The geographer Ptolemy also had the good fortune to study in Alexandria's library. Ptolemy published the biggest compendium yet of information about the spherical Earth, *Geographia.* Ptolemy's work was copied from others, the scrolls and maps seized from passing ships.

Ptolemy also published the *Almagest* about the planets, including star positions taken from Hipparchus. Copying Aristotle's work, Ptolemy made Earth the centre of his cosmology. From Ptolemy until the time of Copernicus most people would believe that Sun, planets and stars circled the Earth.

In some ways this was another understandable assumption. A navigator on Earth need not take into account the distances to stars, for they are too great to affect her calculations. Many astronomers, including this one, keep on the shelf a clear globe with stars depicted on its surface. Though we realise today that this model is just a convenience, for most of history humans believed that stars were truly fixed to an immense sphere.

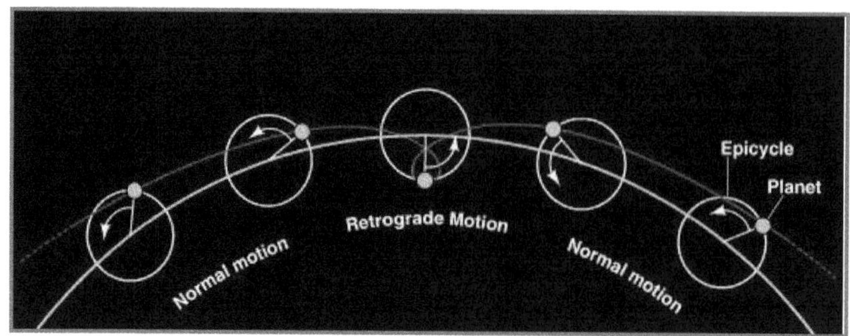

Anyone observing the planets for any length of time will see them appear to reverse course and travel backwards in their paths. To explain this retrograde motion, Ptolemy's cosmology relied upon *epicycles*, spheres within spheres. The planets were each attached to an invisible sphere, in turn attached to bigger spheres surrounding the Earth. The epicycles supported Ptolemy's Universe, preventing it from crashing to Earth.

In the twilight of the Library, one of the most notable and lovely figures at Alexandria was Hypatia. She was daughter of Theon, one of the last scholars at Alexandria. Hypatia was renowned not just for her beauty, but also her accomplishments in mathematics and astronomy. She is said to have translated part of Ptolemy's *Almagest*. Her beauty and influence made others jealous, and Hypatia fell victim to a mob in 415 A.D.

A popular book about the *Cosmos* imagines Hypatia at the Library's burning, vainly trying to stop its destruction. In fact the Library suffered multiple fires, from the siege of Alexandria by Julius Caesar to wars in the 5th century.

The original Library was gone by the time of Hypatia's death. The Temple of Serapis, which housed the last scrolls from the Library's collection, burned in 391 AD.

Ptolemy's cosmology was carefully assembled from observations. As knowledge of Earth composed the earthly portion of the *Geographia,* Ptolemy used the available observations to explain the wider Universe. These observations would have pleased Aristotle. Had the master Plato reviewed Ptolemy's work, he might have complained that this cosmology of many epicycles lacked an overriding principle. There was no single law to explain the motion of heavenly bodies, just epicycles. The world would see the rise and fall of a Roman Empire, the Dark Ages and Renaissance before Ptolemy's cosmology was overturned.

View of Torun from across river

Torun, Poland

The old town of Torun, about 100 miles from the port of Gdansk, looks very much like it did in the time of Nicholas Copernicus. Torun is worth an excursion, for the streets and buildings are a picturesque and well-preserved. The medieval atmosphere is made safer because cars are not allowed on the streets. The tower of the Old Town Hall has an excellent view of Torun and the countryside. Torun even has its own Leaning Tower. You can easily imagine Copernicus in priestly garb watching the sky from church walls.

The writer, who discovered that the Speed of Light is not fixed, was pleased to learn that Nicholas Copernicus, who found that Earth is not fixed, coincidentally was also born on February 19. At 15/17 Copernicus Place you can find the house owned by Copernicus' family, thought to be where he was born in 1473. Today it is a museum and a fine example of a Polish merchant's home. At the Cathedral of St. John the Baptist you can see the font where the infant Nicholas was baptized. A statue of Copernicus stands in front of the Old Town Hall and Nicholas Copernicus University. The University Observatory contains the third largest radio telescope in Europe, and a planetarium.

Many times we will see that the most unlikely of individuals can change the world. For those who knew him, the man who would start a revolutionise astronomy must have seemed quite un-revolutionary. He was forced to leave university after only three years due to lack of funds. Nicholas Copernicus moved through society in the honoured position of a priest. This quiet man of the church led a secret life of science—his nights were spent observing the sky. Quietly he collected and checked his observations, but delayed publishing until the end.

De Revolutionibus Orbium Celestium,
"The Revolution of the Heavenly Spheres" proposed that Earth is not the centre of everything, but circles the Sun. Possibly Copernicus delayed the book for fear of ridicule—people would tell jokes about this Polish astronomer! Copernicus also felt his theory was incomplete, for the planetary orbits could not fit into perfect circles.
The book was finally published in 1543, the day before Copernicus died quietly in bed.

A modest man, Copernicus was buried in an unmarked grave beneath the floor of the cathedral in Frombork, where he had once served as canon. The town, also known as Frauenberg, is about one hour's drive from Gdansk. Only in the first decade of the 21st century were Copernicus' remains discovered and identified; in 2010 he was reburied with full honours. Today in the Frombork cathedral you can see his grave decorated with Copernicus' vision of the solar system.

A statue of Copernicus by Bertelsmann Thorvaldsen is at Krakowskie Przedmiescie in Warsaw, before the Polish Academy of Science. The Copernicus Science Museum, which also houses a planetarium, is several blocks downhill on the banks of the Vistula. The original signed copy of *Die Revolutionibus Orbium Celestium* is kept at the Jagellonian University library in Kraków. Copernicus Crater is on the Moon.

A quiet end was not in store for another man of the church. Giordano Bruno was born 5 years after Copernicus died, trained as a Dominican priest, and embraced Copernicus' new theory. Bruno's own ideas went even further—he speculated that we live in an endless universe, that our solar system is just one of countless others, and that these other solar systems could be home to life! Thinking about other worlds, Giordano Bruno was far ahead of his time.

His seemingly heretical ideas caused Bruno big trouble with authorities. Unable to lead a steady life, Bruno taught and travelled from Italy to Switzerland to England. In Rome, a local official of the Inquisition ordered him executed. Today you can see Giordano Bruno's statue amidst the flower-sellers in Rome's Campo di Fiori, where he was burned at the stake. Revolutionary ideas nearly always meet with opposition.

Statue of Giordano Bruno in Campo De Fiori, Rome

Uraniborg, etching from 1663

Copenhagen, Denmark

In the port of Copenhagen you can see the Tivoli Gardens and the many palaces built for Denmark's royal family. Kronborg Castle, the Elsinore of Shakespeare's *Hamlet*, guards the shore of the Oresund, the strait between Denmark and Sweden. From the Round Tower, an Observatory built in the 17th century, you can see across the Oresund into Sweden.

Copenhagen's most-visited statue is of Hans Christian Andersen's *Little Mermaid.* There is no statue of *The Emperor's New Clothes*, for he had no clothes! The little boy in that story was ridiculed for pointing out the obvious, that the Emperor was stark naked. As with yesterday's epicycles and today's "dark" energy, supposedly sophisticated minds insisted something was there, when it didn't exist.

The castle-like Tycho Brahe Planetarium is dedicated to Denmark's most renowned astronomer. Tycho was born into Denmark's nobility, but his lifelong passion was astronomy. One night in 1572 he saw a new star appear, so bright that it could be seen in daytime. Tycho's supernova was an exploding star, one of the most violent events in the Universe. In its brief moment of glory, a supernova can outshine an entire galaxy. Just as some people thought that Earth or the Speed of Light was fixed, they once thought that the stars were forever fixed in a celestial sphere. This new star seemed to indicate that the sky was not immutable. Today exploding stars are a clue about the Speed of Light.

Denmark's King was so impressed by Tycho that he financed a state-of-the-art observatory for him on the island of Hven, in the Horesund between Zealand and Tycho's home province of Scania. Uraniborg was a castle dedicated to astronomy, surrounded by formal gardens which can be seen today. The observatory was equipped with the finest astronomical instruments that money could buy, though the telescope had yet to be introduced. When the winds of Hven were found to disturb the instruments, Tycho built another observatory underground and called it Stjerneborg.

Tycho Brahe's Observatory

Construction of Uraniborg was said to have consumed about one percent of Denmark's budget. (In comparison, NASA accounts for less than one half a percent of the U.S. budget).

Though the island of Hven and Scania are today part of Sweden, you can visit the remains of Uraniborg. The island can be reached by ferry from Copenhagen or from Landskrona in Sweden. The All Saints Church in Hven contains the Tycho Brahe Museum, with replicas of his astronomical instruments. The museum also maintains Tycho's underground observatory Stjerneborg. The grounds of Uraniborg and its formal gardens are slowly being restored. Today at the Cafe Tycho Brahe, adjacent to his observatory, you can dine on vegetables and herbs grown in Tycho's garden.

Florence, Italy

The centuries when people thought that Earth was fixed were a time of darkness. After the Dark Ages, the Renaissance was a time of light. When you visit Florence, a 90-minute drive from the port of Livorno, you are in the centre of the Italian Renaissance. The Duomo of the Cathedral of Santa Maria del Fiore dominates the skyline as it did in the 16th century. In the Uffizzi Museum, grandest art collection in Italy, you can see masterpieces by Titian, Canaletto, Leonardo Da Vinci and other residents of Florence. When you enter the Uffizzi courtyard from the Piazza Della Signoria, in the corner to your right is the statue of Galileo Galilei.

Born in 1564, Galileo was fascinated by experiments since childhood. Though he was an excellent student, Galileo like Copernicus was forced to leave university for financial reasons before graduating. He found work at the University in Padua, where you can still see the podium he taught from. While teaching in Padua, he preferred performing his own experiments to copying from books.

Alongside the Uffizzi Museum, on the bank of the River Arno in sight of the Ponte Vecchio, is the Museo Galileo. Far smaller than its neighbour, the museum can be experienced in 1-2 hours. After entering, you may head up the stairs to Room I, which is devoted to the Medici collection. The Medici family, rulers of Florence, acquired a huge assortment of astronomical and navigational instruments. Mastery of navigation allowed city-states to extend their influence.

Room II is devoted to astronomy and time. At the room's centre, surrounded by navigational instruments, is a globe dating from 1085 A.D. Room III is dominated by an Armillary sphere. Its many rings and circles track the movements of planets in the sky. With Earth at its centre, the Armillary represents Ptolemy's view of the universe. In the 14th century a translation of Ptolemy's *Geographia* appeared in Italy, popularising his Earth-centred cosmology.

When you enter Room VII, to your right is a case containing Galileo's original telescopes. Through his telescopes Galileo was first to see craters on the Moon, the phases of Venus and the Rings of Saturn. In the same case is the objective lens through which Galileo was first to behold the moons of Jupiter. Though Galileo was initially skeptical of Copernican theory, observations led him to eagerly accept it.

Room VII also contains Galileo's fingers! A century after his death, some of his followers preserved his middle finger for posterity. Today Galileo's finger continually points upward, beckoning us to the sky. The display case also holds two other fingers and a tooth that were found later. In one room you can see not only Galileo's telescopes, but also the fingers that held them.

Galileo's observations were a threat to Ptolemy's system, because they showed objects circling something other than the Earth. As Plato had complained before, some men were afraid of the light. Learned minds of the day refused to peer into his telescope, for fear of upsetting their worldview. Galileo's beliefs led to his trial and sentencing for heresy. He was forced to publicly renounce his beliefs and spend the last ten years of his time on Earth under house arrest.

The 14th century Church of Santa Croce, Italy's largest Franciscan Gothic Church, is a short walk from Museo Galileo. When you enter the nave from the ticket office and turn right, you can see Galileo's grave on the right side near the back door. Within the Church you will also find the graves of Michelangelo and Niccolo Macchiavelli. Persecuted in his time, Galileo is today recognised as one of these great Florentines.

Campo de Miracoli and Leaning Tower

Pisa, once a rival city-state with Florence, is 15 miles from Livorno's port. An ambitious traveler on a day's excursion from Livorno can quickly pass through Pisa on the way to Florence. Galileo was born in Pisa, and grew up in his family home on Via Giuseppe Gusti. You can still see the Galileo family house, now marked by a plaque.

In the 12th and 13th centuries Pisa, with a strategic location near the mouth of the River Arno, controlled the Mediterranean. During this period Pisa built the Campo de Miracoli, Field of Miracles. On the lawn you can see the Baptistery, Campo Santo, and the Leaning Tower. From the Leaning Tower Galileo reportedly experimented with dropping things. He concluded that objects of different mass fall toward Earth at the same rate.

(In 1971 Apollo 15 Astronaut Dave Scott performed Galileo's experiment with a geologist's hammer and a feather on the surface of the Moon.)

Within the Pisa Duomo hangs a bronze fixture called "Galileo's Lamp". A lamp identical to this one supposedly inspired Galileo to think about pendulum motion. Galileo realized that the period of swinging was regular, no matter how far the lamp swung. The lamp started Galileo thinking that a pendulum could be the basis of an accurate clock. In his experiments, Galileo could have used a good clock.

Galileo also suggested testing the Speed of Light. For most of history humans did not know whether light had any speed at all. Aristotle believed that light simply traveled instantaneously. This is an understandable assumption, for light travels so quickly that to observers on Earth its speed seems infinite. Galileo suggested stationing observers with lanterns on distant hilltops to time light's passage, but he still lacked a good clock. A careful scientist, Galileo could only conclude that if light had a speed it was too great for him to measure.

Though today we agree that Earth is not the centre of the Universe, the Copernican model was not complete. The orbits of planets are not perfect circles. The Ptolemaic cosmology could explain this by imagining more epicycles. Scholars kept themselves employed inventing theories with up to 100 epicycles.

The mathematics of epicycles was highly complex, which kept discussion limited to an educated few. These few could point to the non-circular motion of planets and say that Copernicus was wrong. Copernicus even tried to put epicycles in his own theory. Without a guiding principle, the Copernican system was open to criticism.

For most scholars, it was far easier to promote an old system. In addition to curiosity, scientists have the same desires as most other people. One of those desires is to belong to a larger community. Another desire is security, the ability to keep a job and pay the bills. To maintain place and security, it is easier to support the fashionable theory.

A secure life eluded Galileo's friend Johannes Kepler. Though today we know him as a brilliant mathematician, in his lifetime Kepler moved from job to job before becoming Tycho Brahe's last assistant. Tycho's treasure of observations was kept secret from everyone, even his assistant. Only after Tycho's death did Kepler gain access to all the data. Kepler believed in the Copernican system, and sought to find a principle behind it. Like Pythagoras he thought that mathematical relations underlie all of nature.

Fascinated by Plato's geometric forms, Kepler tried to explain orbits as ratios of these shapes. He had greater success with ellipses. The Indian scientist Aryabhatta, who was born in 476 AD, had independently suggested that planetary orbits are elliptical. Starting from the principle of planets orbiting the Sun, Kepler concluded that the orbits are ellipses with the Sun as one focus. Kepler devised laws that described the motion of planets far better than any number of epicycles. Kepler's laws allowed him to predict the transit of Venus in 1631.

Planets in their elliptical tracks are like roller coasters circling the Sun. At the farthest point from the Sun they slow like cars atop a hill. Falling closer to the Sun they gain speed like cars coasting downhill. Nearer the Sun objects orbit faster and farther from the Sun their velocity slows.

After Kepler, the revolution was not complete. There was still no mechanism to explain what held the planets in their elliptical courses. This guiding principle would come from a Law of Gravitation.

NEXT: Moons, Apples and Light

3

Moons, Apples and Light

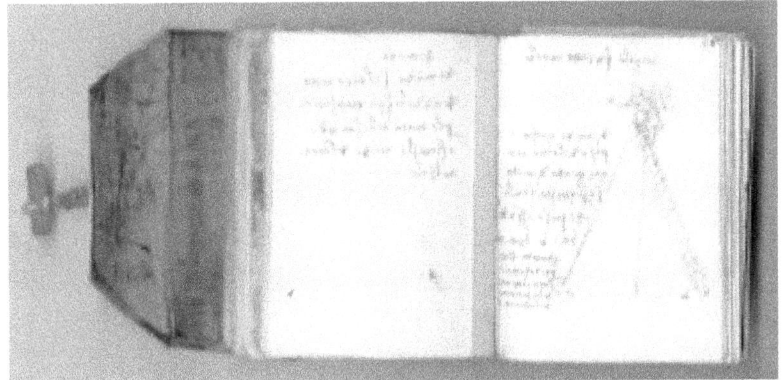

Leonardo Da Vinci, *Man Climbing a Ladder*, c. 1494, displayed at Victoria and Albert Museum 2006-2007.

The hilltop Tuscan town of Vinci, about 25 miles from Florence, named its most famous son Leonardo. The town's 13th century Castello dei Conti Guidi houses part of the Museo Leonardiano, with replicas of Leonardo's inventions. Behind the castle is a wooden sculpture based on Leonardo's Vitruvian Man. The Piazza Della Liberta features a bronze horse base upon Leonardo's designs. Sculptures in the Church of Santa Croce celebrate his baptism. The Piazza Dei Guidi is filled with sculptures inspired by Leonardo's science.

Leonardo Da Vinci birthplace

Leonardo was born 2 miles from Vinci in the tiny village of Anchiano. You can hike there via the Strata Verde, the Green Road that that Leonardo also walked. The modest stone house where he was born is now a museum. The first room you see contains a very old fireplace and a bust of Leonardo. To your left is a darkened room where Leonardo greets you as a hologram. As an artist with a lifelong interest in light, Leonardo would have enjoyed this modern illusion.

The term "Renaissance Man" or woman came to describe someone accomplished in multiple fields. Men and women who make breakthroughs tend to think out of the box by having many interests. A child is naturally curious about everything—it is hard to keep her mind focused on one thing! People working outside their fields—such as astronomers contributing to geology or vice versa, have made many contributions to science.

Leonardo Da Vinci was a true Renaissance man, accomplished as an artist, scientist, architect and inventor. As a master painter, he had a lifelong fascination with light and colour. His detailed drawings of human anatomy and fluid flow were worthy of a modern textbook. His drawings show Leonardo's inventiveness and scientific curiosity.

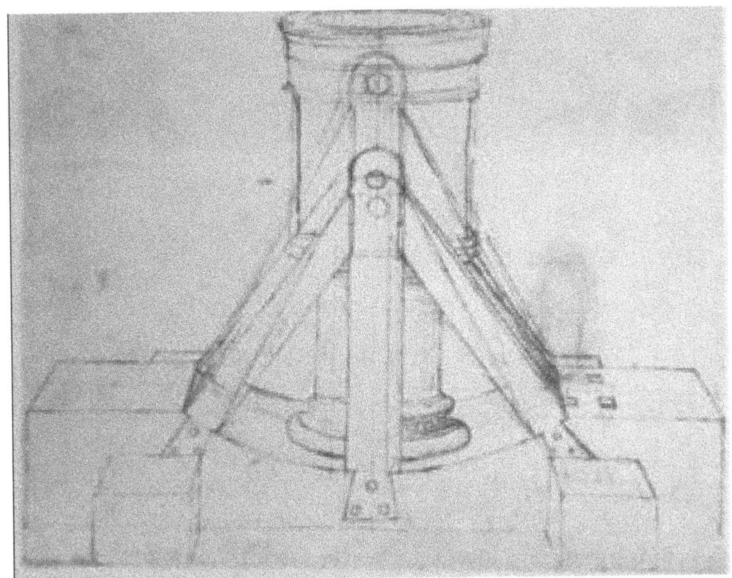

One Leonardo drawing is a study of light rays reflected from a parabolic mirror. On the same page is a tubular device on a moveable mount, possibly a telescope.

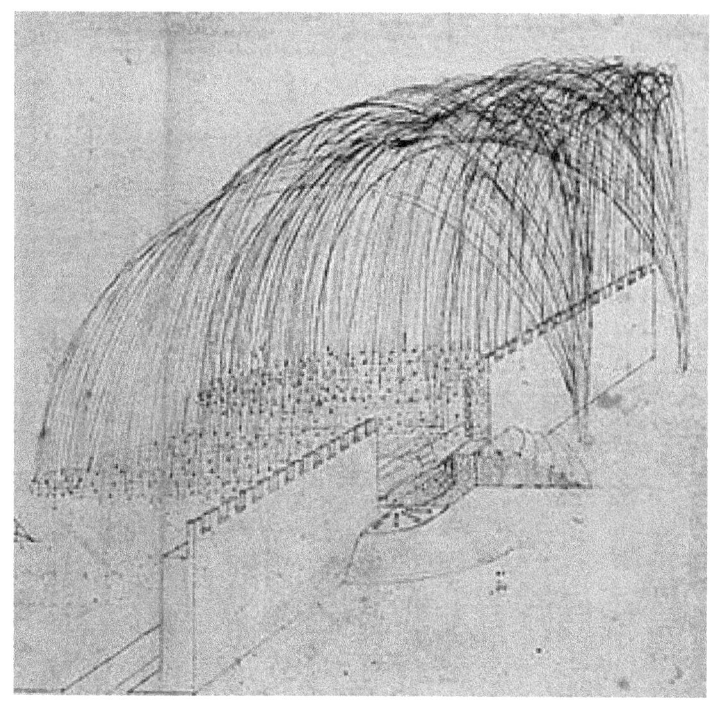

A century before Isaac Newton was born, Leonardo was fascinated by gravity. Among Leonardo's studies of military hardware is a drawing of cannonballs following parabolic arcs, similar to Newton's studies. Leonardo tried to design a perpetual motion machine driven by gravity, blocks and tackles for lifting, parachutes and flying machines for defying gravity.

Leonardo's drawing of a man climbing up a ladder has a vertical line through the man's centre of gravity representing downward force. Leonardo had trained as an artist since childhood, and lacked the training to put his thoughts into numbers. If this line had been extended to Earth's centre, and Leonardo had known more math, he might have developed a theory of gravity.

Newton home in Woolsthorpe. Note window in upper right.

Gravity holds our Universe together, yet we are still learning its secrets. Our knowledge of gravity has advanced in leaps each preceded by an unsolved mystery. Isaac Newton's gravitation solved the mystery of planetary motion, and is still guides spacecraft. Newton's discoveries have implications in the 21st century and beyond.

Isaac Newton was born in 1642, the year Galileo died under house arrest. Though he was initially considered a poor student, Newton showed talent for science at an early age. Like Leonardo he was fascinated by machines, and built a water clock when clocks were still rare.

His uncle had studied at Cambridge, and convinced his family to send the young wizard to Trinity College.

The distance from Woolsthorpe to Cambridge is about 70 miles, an easy drive today but a long journey for a boy in 1661. The ancient halls, gargoyles and stone turrets of the university must have seemed magical to young Isaac. When dining in the Great Hall, Isaac Newton sat at the bottom of the table as a subsizar, the lowest ranking. For his first three years Isaac Newton paid his way through university with menial jobs.

Newton would later write: "If I have seen farther, it is by standing on the shoulders of giants." During Newton's time Cambridge used many teachings of Aristotle, and the Copernican model was still gaining acceptance. In 1664 Newton was finally elected a scholar, but that same year the plague closed down Cambridge. He traveled back to Woolsthotpe, and in two years from 1665 to 1667 started his greatest breakthroughs in gravitation and optics. From humble birth, the plague and solitude, young Isaac Newton produced some of history's greatest discoveries.

Nearly everyone has looked at the Moon, and many have seen apples fall from trees. It took the mind of Newton to see Moon and apples guided by one law of gravitation. Earth and the planets orbit the Sun in Kepler's ellipses. A falling apple is also in an orbit, one interrupted by the Earth. If a hole were dug to allow its passage, the apple would fall like Alice in Wonderland, through Earth's interior and to the other side before falling back and repeating that orbit.

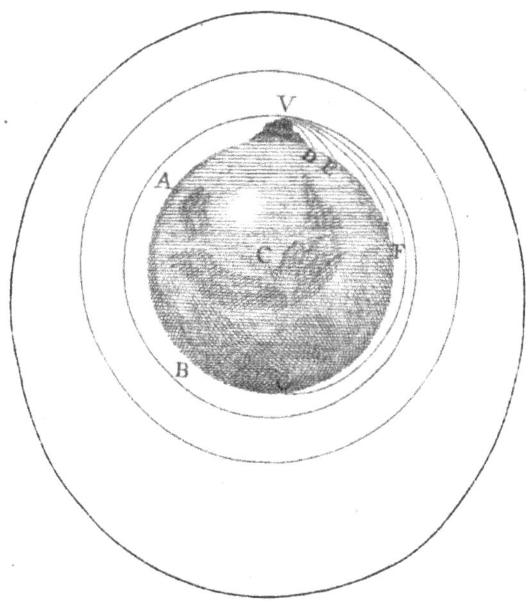

The cover of Newton's book *Principia* is a picture of cannonballs being fired at various velocities, very much like Leonardo's earlier drawing. Newton imagined the cannonball traveling even faster, so fast that it followed a circular path around the Earth without falling back. Three centuries before Sputnik, Newton imagined an artificial satellite. A satellite in a higher orbit would have a smaller velocity. If an object reached an escape velocity, it could leave the Earth completely. So powerful is Newton's Law of gravitation that it still guides spacecraft to the planets.

Where Leonardo had drawn the gravitational force as an arrow leading downward, Newton extended the arrow to Earth's centre. Newton deduced that the gravitational force was proportional to the mass of one object multiplied by mass of the other. Along with his contemporary Robert Hooke, Newton knew that this force was proportional to the inverse-square of the object's distance. To give an answer with the dimensions of force, Newton introduced his gravitational constant *G*.

Newton's Law of Gravitation showed mathematically how the planets move in ellipses. This mechanism completed the Copernican cosmology. Though Aristotle held sway when Newton entered Cambridge, Newton's gravity was the final victory of Copernicus.

Following the publishing of *Principia*, Newton's friend and priest Richard Bentley asked a simple but profound question: If gravity attracts every bit of matter to every other, why doesn't the Universe collapse? Newton, despite his genius, could not answer this question. Much later this same puzzle would lead to Albert Einstein's greatest blunder.

During the same period when plague kept him from Cambridge, Newton also contributed to the study of light. By observing light pass through a prism, he deduced that white light is made up of many colours. This explains why the sky is blue. Shorter, blue wavelengths of light are scattered by Earth's atmosphere. Scattering of blue light makes the daytime sky appear blue.

Though Newton believed that light is composed of particles that he called *corpuscles*, diffraction showed him that it also behaved like waves. Argument over wave vs. particle would rage for centuries. Redshift of light spectra would someday signal expansion of the Universe.

On a trip to Woolsthorpe you can see not just Newton's birthplace, but also the window through which light entered his thoughts. Through the upper right window in the photo, Newton studied the refraction of light. You can easily imagine Newton holding a prism in this light to see it refracted into colours. You can also see an apple tree like the one that supposedly inspired Newton. Cambridge University planted an apple tree to honour Newton, but his greatest discoveries belong to his humble home of Woolsthorpe.

As this new edition of the book is being written, we must shine light on Newton's contemporary Robert Hooke. This great experimenter also came from humble beginnings, losing his father at age 13. Hooke grew up to attend Westminster School and Oxford University, then became assistant to experimentalist Robert Boyle. Starting in 1660 Boyle and Hooke formed the Royal Society, and Hooke became the Society's first Curator of Experiments. He helped discover Boyle's Law of gases, kept track of time by building a spring-powered watch, and is remembered in the famous Hooke's Law for springs.

Hooke helped rebuild London after the great fire of 1666, becoming assistant to the architect Christopher Wren. He contributed to the design of the Royal Greenwich

Observatory and even St. Paul's Cathedral. Hooke made contributions to geology and paleontology, wondering about the origins of earthquakes and fossils. Robert Hooke was such a Renaissance Man that some people call him England's Leonardo.

Hooke also observed light being reflected in a prism, and believed it was made of waves. With his knowledge of light Hooke built a new type of astronomical instrument, the Gregorian Telescope. Like Galileo he observed craters on the Moon and the Rings of Saturn. Hooke tried to use parallax to measure the distance to stars. In addition to looking up at the stars, Hooke looked down into a microscope.

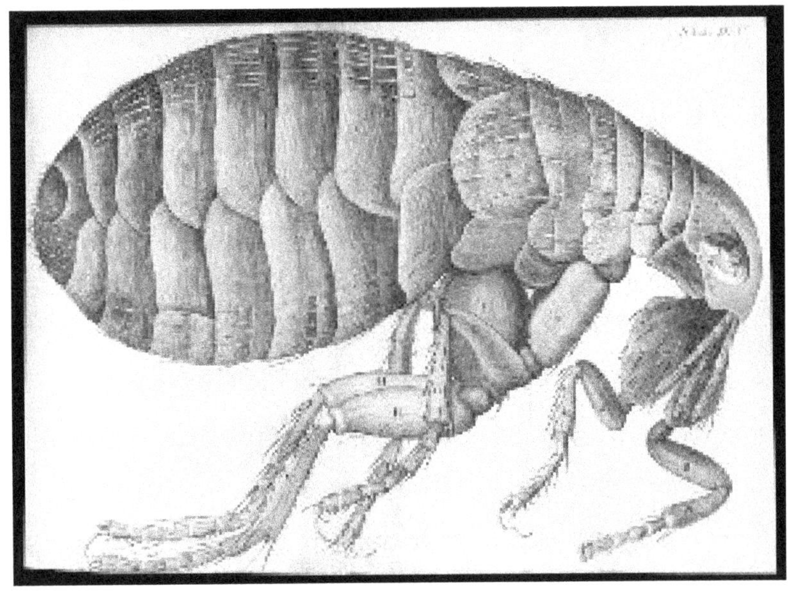

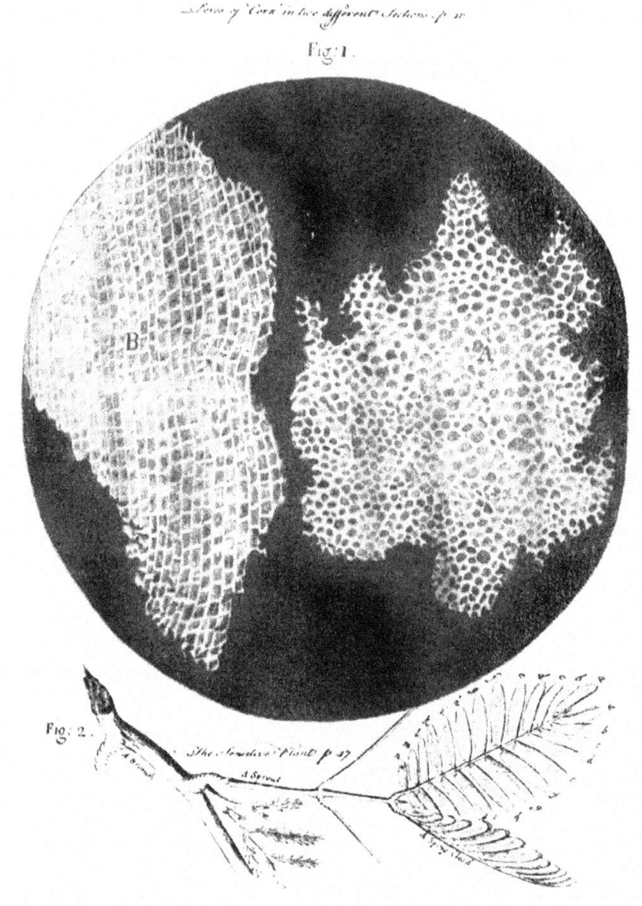

Hooke may be best known for publishing *Micrographia.* This book, which became a science bestseller, was filled with Hooke's drawings of the microscopic world. He discovered that life is divided into tiny *cells,* similar to the cells in a beehive or a nunnery. Why our life is made of cells, and why they are so tiny, was a mystery for centuries. The last chapter of this book explores the possible answer, and The Speed of Light.

On gravity, Hooke first suggested that it obeyed an inverse-square law, as Newton himself acknowledged. This led to rivalry over who discovered gravity, along with the disagreement about whether light forms waves or particles. Though Newton is remembered as a great scientist of his time, Robert Hooke also made many contributions to our knowledge. Today a microscope used by Robert Hooke has traveled to America, the National Museum of Health and Medicine in Silver Spring, Maryland.

Le Observatoire, Paris, France

(In October 2019 I had the privilege of lecturing about the Speed of Light to an international conference at the Paris Observatory and Institute d'Astrophysique.)

The Speed of Light was found in the City of Light. The Paris Observatory is at 61 Avenue de l'Observatoire, a short distance from the Latin Quarter. The nearest Metro station is at Denfert-Rochereau, but a more scenic walk begins at Jardin du Luxembourg. From the gardens you may walk South along the Avenue past the Fontaine de l'Observatoire and a statue of four women holding the world up. Institute d'Astrophysique, where I spoke, is next door.

Our Sun King Louis XIV financed the Observatory, wishing to stay ahead of rival nations. On the floor in the Observatory you can see a line marking Zero degrees longitude. Britain's Royal Greenwich Observatory has a more famous line also marking the Prime Meridian. Location of Zero degrees longitude is arbitrary, so maps drawn in different nations would place it in their territory.

Ptolemy's *Geographia* placed Zero on the island of Rhodes and the United States measured longitude from Washington's Naval Observatory. As France and Britain competed to build empires, they also competed for the honour of having Zero in their territory. Eventually the influence of Britain's sea power led to Greenwich being adopted by other nations, though the French disagreed. Observatories and astronomy were very important for competing nations.

Galileo had tried to measure the Speed of Light with lanterns on a distant hilltop, but lacked a good clock or a distant hill. The first evidence that light had a speed came from another Galileo discovery, the moons of Jupiter. When a young astronomer named Ole Roemer arrived at the Observatory in 1672, there was an anomaly in observations of the innermost moon Io--it appeared from behind Jupiter late. Roemer realized that if Io's moonlight had a finite speed, the moon's appearance would be delayed.

Using data from the observatory Roemer was first to measure the Speed of Light, though his estimate of 210,000 kilometres per second was 30% too slow. After years of work, in 1675 Roemer was bold enough to present the results. He also predicted that on November 9, 1676 Io would appear at 5:35:45 rather than 5:25:45 as astronomers had calculated. On that date Io emerged from behind Jupiter 10 minutes late, precisely as Roemer had predicted.

Having made a great discovery, Roemer was unable to convince his elders. The Observatory's Director, Domenico Cassini was a distinguished astronomer whose name would be given to a spacecraft orbiting Saturn, but he believed that light traveled instantaneously. Though Roemer was correct, Cassini and others insisted that there was no Speed of Light. 50 years would pass before other experiments verified that light had a speed.

James Bradley in 1728 used stellar aberration, when the position of stars appears to change because of Earth's orbit around the Sun. French physicists Hippolyte Fizeau in 1849 and Leon Foucault in 1862 used a system of rotating disks on distant hilltops to measure the Speed of Light. Since Roemer's time, many other experiments have been devised to accurately measure *c*. Why light travels at 299,792.5 kilometers per second, not faster or slower, remained a mystery.

Isaac Newton could contemplate the sky, imagine cannonballs traveling at orbital speed, and calculate that a satellite in a circular orbit would have a certain velocity at a given altitude. Roemer visited Newton in 1679; very likely they discussed the Speed of Light. Since Newton knew that Moons and apples are both guided by one law, and believed that light is made of particles, might he have suspected that light is also affected by gravity?

One consequence of the Speed of Light has been *Black Holes*, objects so dense that not even light can escape. In 1783, the year the Montgolfier Brothers launched the first manned balloon, a priest named John Michell deduced the existence of Black Holes from Newton's Laws. By equating the Speed of Light with Newton's escape velocity, Michell concluded that an object with enough mass would trap even light in its gravity. The term "Black Hole" would not enter use until the 1970's.

Mathematician Pierre Laplace promoted the idea of what we call Black Holes in a book published in 1796. Interested in many things, Laplace also suggested that the Solar System condensed from a cloud of gas. Of all denizens of the Universe, Black Holes are among the most mysterious and powerful. As we shall see in this book, they may even be related to the formation of the Solar System.

Thomas Young was another polymath, a man with many interests. Among his many achievements, he translated Egypt's Rosetta Stone into English. As a physician, he studied the human eye and concluded that it processes three colours of red, blue and yellow. The eye's structure may have given Young ideas about light itself. In a talk delivered in 1801, Young argued that light takes the form of waves. This appeared to contradict Isaac Newton, who favoured a particle theory of light. Since Newton was Young's wave theory of light was not accepted for decades.

Maxwell, Light and the Hypothetical Ether

For most of history communication was limited by the speed of human travel. Starting in the 1840's the electric telegraph was introduced. During the 1870's Alexander Graham Bell would perfect the telephone, allowing voices to travel in electric wires. In the 19th century communications began moving at the Speed of Light.

Much of our familiarity with light comes from the work of James Clerk Maxwell. The eldest son of a wealthy Scottish family, Maxwell's background started different from Newton's. Like Newton, Maxwell started his best-known work while an undergraduate at Trinity College. His other interests included the Rings of Saturn and the kinetic theory of gases. Fascinated by light, he also made the first permanent colour photo!

Maxwell built upon the experiments of Michael Faraday with magnetic fields, and the discoveries of Andre-Marie Ampere about electric current. Maxwell showed that electric and magnetic fields were both part of one phenomenon, *electromagnetism*. Electromagnetic waves can take the form of infrared, microwave, radio, UV, X-rays, gamma rays or visible light. Combining Faraday and Ampere's Laws with some equations of his own, Maxwell produced a set of equations describing all electromagnetic interactions.

Maxwell's equations allowed calculation of the speed of these waves. That speed of *310,000 km/sec* was almost exactly the 19th century value of the Speed of Light. Maxwell thought this too close for coincidence--he proposed that light was also a form of electromagnetic waves. The same Maxwell equations described the propagation of light, and showed that light can indeed take the form of waves.

Our lives begin in a tiny womb with no visible light— but we are not isolated from light's influence. The warmth we feel from our mothers is infrared radiation, another form of electromagnetism. When we feel heat, we feel a form of light stretched into long wavelengths. Certain animals, such as snakes, have eyes that see in the infrared. When we are drawn to warmth of a fire or another person, it is another way of seeking light.

There has long been speculation whether light's speed has always been the same. In 1874, shortly after Maxwell published his equations, William Thomson claimed that the Speed of Light was slowing. Thomson was one of the most prominent scientists of the late 19th century, honoured as First Lord Kelvin and via the Kelvin temperature scale. Along with William Tait, Thomson published a paper claiming that light slows down. Though the subject is not often discussed, many scientists have sought a change in *c*.

After Maxwell, the debate remained whether light was particle or wave. Newton had believed that light was made of particles called corpuscles, though his own experiments showed that it had wavelike characteristics. Maxwell's discovery that light is an electromagnetic wave appeared to settle the argument. The advent of quantum mechanics would show that they were both right.

Wavelike properties led to one of the biggest misconceptions about light. Since sound waves travel through air and ocean waves through water, it seemed logical that light also propagates through some sort of medium. This imaginary substance was called *ether*. Nearly every scientist of the 19th century, including Maxwell himself, believed in ether.

Since light travels throughout the Universe, this ether was assumed to fill all of Space. Ether shared with epicycles all the hallmarks of a scientific misconception. Since we can't see ether in the blackness of Space, it was assumed to be invisible, like the epicycles. Ether's existence was only inferred--no experimenter claimed to have a test tube of it or even a microscope slide.

Earth was believed to travel through the ether like a ship through water. The concept of ether was dealt a mortal blow by the experiments of Albert Michelson and Edward Morley. Both men were fascinated by the Speed of Light. If light traveled through the ether at fixed speed and Earth moved through the ether, then the speeds of light parallel and perpendicular to Earth's motion should have been different.

Michelson and Morley's experiment involved an interferometer with two perpendicular arms. The differing speeds of light down the two arms would have created interference patterns. To everyone's surprise, the experiment detected no such interference. Even after many trials, Michelson and Morley failed to find any indication of Earth's motion through ether. The Speed of Light appeared the same in all directions.

In the latter part of the 19th century, may theorists attempted to explain Michelson and Morley's results. Hendrik Lorentz introduced the mathematics that would lead to a solution, mathematical equations for length and time. This Lorentz Transformation suggested that lengths of objects are contracted in the direction of their velocity. In Lorentz's interpretation the Michelson--Morley apparatus literally contracted, making *c* appear the same in all directions.

During 1889, year of the Paris Exhibition, another great event occurred in the history of light--Vincent Van Gogh completed *Starry Night.* Van Gogh was living in the Saint Paul-de-Mausole asylum in Provence after the infamous incident with his ear, suffering from depression and epilepsy. During his time in the asylum Vincent Van Gogh produced some of his greatest paintings, including *Irises* and this self-portrait. As Newton had seen that light in his window was made up of colours, Van Gogh looked out the asylum window and saw *Starry Night.*

In *Starry Night* the Universe is not full of darkness and the stars are not just points but colourful swirls of matter. In the centre is a spiral like a galaxy--at the time no one knew whether “spiral nebulae” in telescopes were objects within our Milky Way or other galaxies. Van Gogh painted Space as being filled with light, as if invisible matter filled the space between stars. His vision of *Starry Night* was a century ahead of his time.

Despite his genius in capturing light, Van Gogh sold almost no paintings. While other artists of the time found recognition and fortune, Vincent lived in poverty. The Paris Exhibition of 1889 lit up the city, but showed no works by Van Gogh. Science and art are alike in that geniuses are often not recognized in their time.

The rooms where Van Gogh painted *Starry Night* are at the Monastery of Saint Paul de Mausole in Saint Remy, Provence, and have been turned into a museum.

Today the Speed of Light has been measured:

299,782.6 m/sec

Why it is that value, not faster or slower, was a great mystery.

As Newton's theory of gravitation solved the problem of planetary motions, his studies of light led to other puzzles. Maxwell's equations showed that light was a form of electromagnetic radiation with a velocity c. The assumption that light waves traveled through an invisible ether was contradicted by Michelson and Morley's experiment. Lorentz's theory that objects contracted with velocity was a step toward a solution. The principle that would explain this puzzle would not come from any professor or doctor of science. It would come from a clerk in the Swiss patent office.

Next: If You Are Within the Sound of My Voice

4

If You Are Within...

That light requires the same time to travel from A to M is neither a supposition nor a hypothesis about the physical nature of light but a stipulation which I can make at my own free discretion— Albert Einstein

Statue of Albert Einstein, Washington DC.

Today Albert Einstein's statue occupies a place of honour in front of the National Academy of Sciences, not far from the National Mall. Albert Einstein was another unlikely suspect to start a revolution. Like Newton, he would at times be considered a poor learner.

When he was two years old, while Michelson and Morley were first performing their experiment, Albert had not learned to speak at all. The very act of putting his thoughts into words was a challenge for him. Growing up in a world of Prussian militarism, Albert grew to dislike the regimen of school. In his studies he preferred marching to his own drummer.

Like Newton and Da Vinci, Albert developed a love for gadgets. Possibly he inherited this fascination from his father, who ran an electronics business with Albert's uncle. When Albert was 5 his father gave him a magnetic compass; the movements of the needle in response to unseen forces began a lifelong fascination. In later years Einstein would call Earth's magnetic field one of the greatest unsolved problems of science.

Though Albert disliked sports and games, he led a rich life outside of school. The dynamos and electrical devices built by his father were sparking, whirring magic to a boy. His uncle Max Talmud gave him a book on algebra while Albert was still in elementary school. When Albert was 12 his uncle gave him a book on geometry.

Einstein loved his "sacred little geometry book" as a child would a treasured plaything. He was inspired by Euclid and Pythagoras' search for harmony. His uncle gave him Immanuel Kant's *Critique of Pure Reason* when Albert was 13. When his school finally taught him algebra, Einstein was at least 2 years ahead of the class.

When Albert was 14 the business run by his father and uncle fell on hard times and was forced to close, and the family moved to Milan in search of other work. Albert had grown up with this business, and the loss affected him deeply. Later his science career would take him from Milan to Bern and eventually Princeton, uprooted from a sense of home. He learned to think independently, insulating himself from the fashionable ideas of his time.

At university, Einstein preferred his own studies to the assignments. He had a reputation for borrowing the textbook on the night before the exam to study. In his last year he ran afoul of the professor of experimental physics. Einstein's original suggestion of a way to test Michelson-Morley did not impress the professor, whose name is forgotten.

In 1900, academic success depended on pleasing a large number of the old and inflexible. The corollary is that the disfavour of one can cripple an academic career. Though his fellow students all received recommendations for further positions, Einstein did not. The great scientist could not find a job in science.

With the help of family and friends, Einstein finally found a position at the Swiss Patent Office. Examining patent applications proved to be satisfying work for someone who had grown up with electronics. In this period, separated from the usual circle of academic discourse, Einstein produced his four great papers of 1905. Any one of those papers would be considered a huge breakthrough in physics. Through his trials, Einstein retained the sense of wonder that children share.

Bern's best-known landmark is the Zytglogge, a clock tower that was once the city's western gate. The colourful astronomical clock was built in the 16th century. At four minutes before each hour a parade of mechanical characters begins at the clock. From 1903 to 1905 Einstein lived up the street at Kramgasse 49, and saw this parade many times. Einsteinhaus is now a museum where you can see his writing desk and other mementoes.

The turn of the 20th century was another exciting time for physics, the beginnings of a quantum revolution. This had begun with another mystery of light. A heated body produces radiation in a characteristic spectrum, like a hot piece of iron glowing red. At very low wavelengths this radiation seemed to vanish. This "ultraviolet catastrophe" defied Maxwell's equations. Max Planck spent years trying to solve this mystery. Finally in 1900 he solved the catastrophe with an "act of desperation," introducing the quantum value *h*. Multiplying by the Speed of Light *c*, the energy of light is quantized into tiny amounts by *hc*.

As a conservative man of science, Planck tried everything possible to avoid introducing *h*. After 6 years of work he could find no other way to explain the data. The Planck formula for blackbody radiation fit the data curve precisely. When a prediction line matches the data, it probably means something.

Quantum mechanics predicts that the tiniest particles cannot be located precisely. Such measurements are subject to the Uncertainty Principle. While we cannot locate a particle precisely, we can predict the probability that a particle will be in a given location. With Quantum mechanics, predicting the behaviour of particles becomes a game of statistics. Nuclear energy, transistors and lasers, some of the foundations of modern technology, are a result of quantum mechanics.

Quantum mechanics predicts that Space can be filled with "virtual particles." Paul Dirac used QM to predict the existence of *antimatter,* a negative counterpart to matter. Statistics predict that under certain conditions, pairs of matter and antimatter particles can appear in a process called "pair production." Quantum mechanics helps explain how the matter that we are made of formed.

Einstein's first 1905 paper concerned the photoelectric effect. Light shining upon metal produces electricity, an effect that powers solar cells today. The behaviour of the power produced could not be explained by Maxwell, or by wave theories of light. Einstein showed that this effect could be explained if light were quantized by *hc* into particles called photons. This returned to Newton's belief that light takes the form of particles.

New ideas often have difficulty even getting published. In Einstein's time there was no "peer review," papers only had to be approved by an editor. Fortunately for the history of science, the editor of *Annalen der Physik* was Max Planck. If not for Planck's interest, Einstein's papers might have been delayed indefinitely.

Max Planck recognised the importance of Einstein's papers, and accepted them for the journal. After their publication, the reaction was still a deafening silence. Though Einstein was fortunate that his 1905 papers were promptly published, not until 1911 did the magazine

Scientific American mention his work. Einstein would wait years before the wider world recognized his achievement.

Something in Albert's upbringing made him think different. He gained curiosity and love of science from his father and uncles. Somewhere in his journeys, he came upon a truth that children can learn and adults can forget. *Space and Time are one phenomenon.*

Humans are born hardwired to think this. In Galileo or Newton's time the distance between towns would be given in days traveled rather than miles. Today when you ask most people the distance from work to home, very few know the distance in kilometres but they nearly always know the time between those points, often to the minutes. In an age of clocks, this shorthand keeps us from being late to school.

The conversion factor between distance and time is the speed of our vehicles. When we are hurrying to work, our speed is usually the speed limit of the road. Nature has her own speed limit, built into everything from the tiniest atoms to the very Universe, known as the Speed of Light. This value c is not just the light's speed; it is the speed of any massless particle and the speed of information. It was originally named for the Latin *celeritas*, but more accurately it is a *conversion* factor between Space/Time. Whether conversion is expressed in miles per hour or meters per second, it is a velocity.

The start of the twentieth century also saw the introduction of radio. The young inventor Marconi had demonstrated practical radios in 1897. By the end of 1901 Marconi had sent radio signals across the Atlantic. The electromagnetic waves of radios, telephones and televisions respectfully obey Maxwell's equations.

Nearly all of us have spent hours listening to a radio. Disk jockeys often begin an announcement: "If you are within the sound of my voice…" In this phrase lay the truth that Einstein would struggle to put into words: Space and Time are one phenomenon related by *c*.

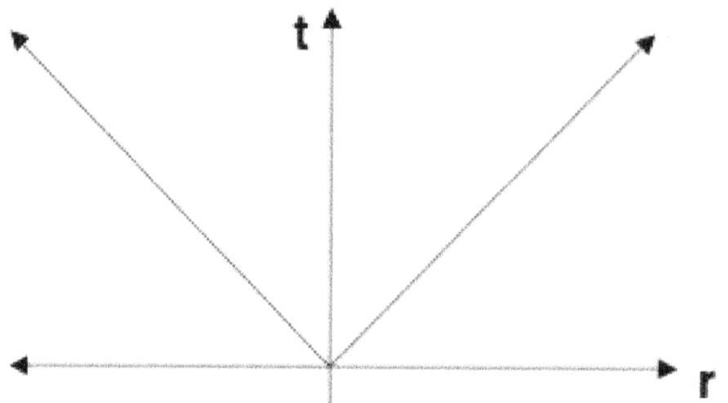

This can be expressed in this 2-dimensional page if the three spatial dimensions x, y, and z are compressed into one dimension, r. Now we can put the additional dimension of time into the vertical. At the centre is a radio station, broadcasting a signal at the Speed of Light. Seen in this drawing, the spreading of these electromagnetic waves appears as a cone, a light cone.

Events elsewhere in Space/Time can be either outside or inside this cone. An event on the Moon less than one second after our signal transmits is outside the cone. Since light takes more than one second to reach the Moon, our radio signal does not reach this event in time.
The separation between these two events is said to be *Spacelike.*

If one second after our signal is transmitted, another rocket is launched from a point on Earth, that event is inside the light cone. Since it takes light less than a second to cross the Earth, our signal reaches this nearer rocket before it launches. The separation between these events is called *Timelike.* Their separation is literally a matter of time.

If the centre is a baby's birth, the cone represents her family broadcasting and tweeting the happy event. The expanding light cone represents the events that the baby's birth will influence. The events that influence a baby's birth are also contained in a cone, converging at the birth. The event is at the apex of both a converging and an expanding cone of light.

Mathematician Johann Lambert in the 18th century found formulae for areas of triangles in curved spaces. He found that when applying those formulae to spherical spaces, the radius of those spheres was the square root of a negative number. These so-called imaginary numbers would turn out to be critical in Space/Time. The Timelike separation of an event within the light cone also turns out to be the square root of a negative number. For a spherical Space, mathematics favours a fourth dimension of Time.

To put Space and Time on the same page, we need a conversion factor. That factor shows us how much vertical Time to draw for each unit of horizontal Space. The conversion is provided by c. The value c is much more than just the Speed of Light, for the rules of Relativity apply even in a dark room with the lights out. Even when no one can see you, you can't exceed the Speed of Light.

From this principle, called *causality*, one can derive all the equations of Special Relativity. Mathematician E.C. Zeeman in 1964 showed that the causality principle led to the Lorentz equations of Relativity. The causality principle does not require the Speed of Light to be constant. Albert Einstein would later experiment with a varying c.

Seen in our 3 dimensions of Space and one of Time, the light cone looks like the famous picture of a radio transmitter. Marconi's experiments showed that an antenna works best when mounted on a vertical tower. Signals travel outward at the Speed of Light c. The expanding spherical waves of the radio transmitter are the light cone in four dimensions. Outside the spherical wave of a signal, a listener has Spacelike separation. Inside the expanding sphere, the separation is Timelike, within the sound of the announcer's voice.

Michelson and Morley's experiment appeared to show the same Speed of Light in every direction. Lorentz found mathematical equations to account for this, but could not find a principle behind them. Einstein added a simple stipulation: Light's speed *c* is constant regardless of location. Einstein himself did not consider this to be a theory about light. He wrote that a constant Speed of Light was "*Neither a supposition nor a hypothesis about the physical nature of light but a stipulation which I can make at my own free discretion.*"

The predictions of special Relativity would fill physics books for a century. As Lorentz had first suspected, objects traveling at high velocity literally contracted in length. The very passage of time would slow relative to stationary objects. A starship accelerated to near-light velocities could reach another solar system in a short time by the ship's clocks. Upon its return the crew would find that decades or centuries had passed back home. Since Einstein's time, many experiments have verified these predictions. As spaceflight became a reality, more of these effects of Relativity would be detected.

Relativity was not yet complete, as Einstein himself realised. It made no allowance whatsoever for gravity. Einstein would consider this problem for 10 more years. For several more years he would retain his patent office job, finally moving to a teaching position in 1909. Around 1907 he had a vision similar to Newton's, and of equal importance. "For an observer falling freely from the roof of a house there exists at least in his immediate surroundings no gravitational field," he would write. This was Einstein's Principle of Equivalence. He realized that an object in freefall feels weightless, an effect we later witnessed in orbiting astronauts. Einstein called this the most fortunate thought of his life.

In 1911 Einstein spent a lot of time on a theory with a varying Speed of Light. By this theory, c would change in the presence of gravitational fields, causing the paths of light rays to bend. Some people claim that Special Relativity required c to be fixed, but Einstein himself was willing to consider that it changed. Though his Special Relativity stipulated that light's speed is the same regardless of location, Einstein did not forbid it to change over time.

Inspiration in physics requires the perspiration of mathematics. Einstein was by his own admission not the best mathematician, so he sought help from friends. Marcel Grossman helped him with Riemannian geometry. Contrary to the geometry that Einstein grew up with, this was non-Euclidean. Riemann's vectors and tensors allowed for spaces that were curved rather than flat. Using Riemannian geometry, Einstein in 1915 published his General Theory of Relativity.

General Relativity stated that Space/Time is not flat, but is curved by the presence of mass. Any massive object causes Space/Time to be warped, as a stone placed on a rubber sheet causes the sheet to dimple. A marble rolling at high speed past this gravity well will see its course slightly altered. If the marble travels too close and too slow, it will fall into the stone's well. Given the right velocity, the marble will circle around the well as a satellite. Circling at a greater distance from the stone requires a lower velocity. Gravity was not a force, but the very curvature of Space/Time. Since light travels through Space/Time, General Relativity predicted that light itself would be bent by gravity.

Einstein's General Relativity solved a longstanding problem concerning Mercury's orbit. The innermost planet orbits in a Johannes Kepler ellipse with an axis that precesses at *5,600 arc seconds* per century. Many other objects exert small influences upon this orbit, but when all the influences were added there was a discrepancy of *43 arc seconds* per century. In the latter part of the 19th century scientists tried many theories to account for the discrepancy.

Again scientists hypothesized invisible forces. At one point an unknown planet called Vulcan was inferred to exist within Mercury's orbit; today it is the name of Mr. Spock's *Star Trek* home. The curved Space/Time of General Relativity explained this mystery without imaginary forces. When his prediction matched the data Einstein was giddy with joy.

Special Relativity used the equations that Lorentz had already introduced, and may be considered as an idea whose time had come. General Relativity was a huge step forward, a revolution in our understanding of gravity. Einstein did not set out to prove Newton wrong, rather he ensured that his equations reduced to Newtonian dynamics in general cases. By going beyond Newton, General Relativity allowed Einstein to envision the very structure of Space/Time.

Though Special and General Relativity are seen as two parts, they were never truly linked. Special Relativity still made no allowance for gravity. Reducing the equations of General Relativity to those of Special Relativity proved to be a daunting challenge. To link the two parts, one had to picture the shape that gravity would bend Space/Time into. Einstein dared to imagine the entire Universe.

Next: Einstein's Sphere of Light

5

Einstein's Sphere of Light

Nature is an infinite sphere...—Blaise Pascal

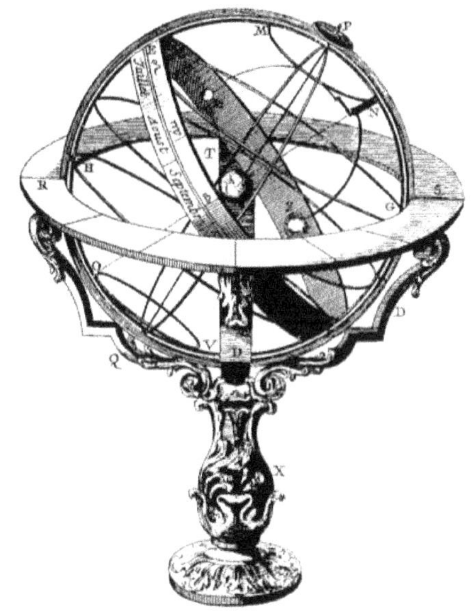

A book that everyone should read was titled *Relativity: The Special and General Theory* by Albert Einstein. The 1961 edition is subtitled *A Simple Explanation That Anyone Can Understand.* The book is available in paperback for less than 10 dollars, far less than overpriced textbooks. In his simple explanation Einstein uses concepts, like imaginary time, that are not taught in schools today. The shortcuts used in textbooks written by others hide the true beauty of Einstein's Relativity. No one else understood the subject like Einstein himself.

In Chapter 31 Einstein attempts to imagine the entire Universe! Like Pythagoras 2500 years before, Einstein is motivated by a search for harmony. He follows what would be called the Cosmological Principle: By this principle the Universe looks the same no matter what direction one looks, and every bit resembles every other bit. He rejects a flat Universe, for his General Relativity shows that Space/Time is curved. A rebellious soul, he rejects boundaries and considers the Universe "finite yet unbounded". The obvious example is a sphere.

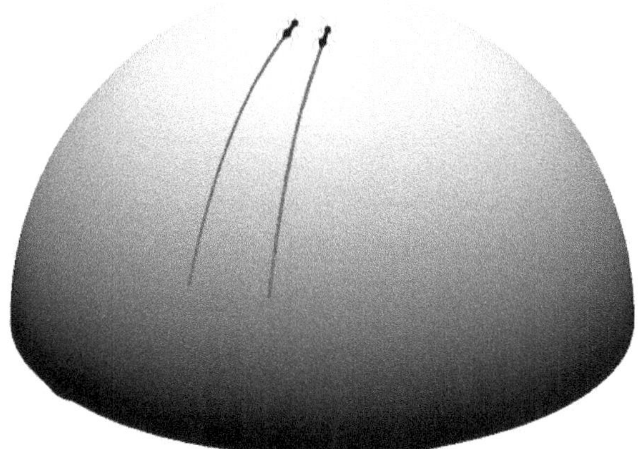

An ant on the surface of a sphere lives in a Universe that is finite yet unbounded. If the ant set off in a line of any direction she would eventually end up at the point where she started. From the ant's perspective, a large enough sphere would seem flat. That is an understandable assumption, for an ant's world is small. Einstein's spherical Universe is also closed yet without boundaries.

Einstein's General Relativity predicted that mass causes Space to be curved. If the Universe contained enough mass, it would be curved into a sphere of four dimensions, rather than three. The three dimensions that we occupy would be confined to its surface. Light traveling in any direction around this sphere would follow a circular path, like a satellite in orbit.

Blaise Pascal, who lived in Paris, was one of the great mathematicians of all time. As a teenager he published his own mathematical theorems and designed a mechanical calculator. In the 17th century Pascal wrote,

"Nature is an infinite sphere of which the center is everywhere and the circumference nowhere."
Pascal also thought that the Universe was spherical.

In the 'hood of Baltimore, in front of a housing project, sits an unassuming little house. The mind that lived within these brick walls gave us *The Raven, The Fall of the House of Usher*, and *The Pit and the Pendulum.* 203 Amity Street was the home of Edgar Allan Poe. While physicists see the pendulum as a simple harmonic oscillator, Poe saw the terrifying, inexorable approach of death.

After *The Raven* brought fame and success, the heart of Poe's beloved wife stopped beating in 1847. Poe himself would pass on just 2 years later, living only slightly longer than Blaise Pascal. Possibly sensing the pendulum's approach, Poe in 1848 finished his favourite work. Of all his writings, poetry and prose, Edgar Allan Poe was most proud of *Eureka*. It was a poem not of death or suffering, but about our Universe.

Like Pythagoras and Pascal, Poe thought a sphere the most natural shape. Most prescient, Poe suggested that our giant Universe had expanded from a single tiny point! "From one particle, as a center," Poe wrote, "let us suppose to be irradiated spherically—in all directions—to immeasurable but still to definite distances in the previously vacant space." He thought that our galaxy was just one of many. The master of dark fiction also predicted that the Milky Way's centre contained a massive unseen object, which he called a "non-luminous sun." Predicting a Black Hole at the centre of our galaxy, Poe's *Eureka* was far ahead of his time.

Albert Einstein concluded that we live in such a sphere. The local conditions of Special Relativity, which assume a flat Space with no gravity, apply in our small patch of the Universe. If a starship could travel much faster than light, speeding off in any direction would eventually lead back to where it started. Speeding off in an entirely different direction would return it to the same point. (In an episode of *Star Trek: The Next Generation* the crew of the Starship *Enterprise* find themselves trapped in a spherical space. They leave a buoy behind to mark their starting point. Any direction the *Enterprise* warps to returns them to the buoy.)

Though curved on the large scale, for small distances Einstein's Universe may be considered flat. From tiny raindrops to planets, large objects under the influence of gravity tend to form spheres. The Universe is very large. Since long before Columbus, the "Flat vs. Curved" debate has raged on Earth. Everyone knows which side wins.

The Universe imagined by Einstein has no centre, but every bit does feel an attraction from the whole. Newton's calculus had shown 250 years before that the gravitational attraction from a spherical shell is the same as is all the mass were in the centre. The combined gravity of the Universe would pull each point inward. Newton's mathematical proof applies in 3 or 4 dimensions.

Here Einstein encountered a conflict. The same gravity that curved Space/Time into a sphere would cause that Universe to collapse. (I must add here: *Gravity would cause a spherical space to collapse...unless it expanded.*) The momentum of expansion would prevent the Universe from collapsing. If Einstein had proposed an expanding Universe, it would have been one of the great scientific predictions of all time. At first others would have ignored this prediction, for astronomers took 15 years to discover that the Universe expanded.

"A galaxy far, far away" is the opening of *Star Wars* movies, but in 1917 our Milky Way was the known Universe. Astronomers did not agree how far its disk of stars extended, where its centre was located, or even if the

Milky Way had a centre at all. They disagreed whether "spiral nebulae" seen in telescopes were part of the Milky Way or other galaxies. Though the spectra of these objects showed a shift into red, other data seemed to show that the Universe was unchanging.

Einstein hoped to satisfy a Perfect Cosmological Principle, where the Universe looked the same everywhere in both Time and Space. Possibly motivated by a search for harmony, Einstein believed his sphere to be static and unchanging. In a static Universe, the Speed of Light would be constant. A spherical balloon would collapse without something to support it. Something had to prevent the static universe from collapsing.

Here Einstein made his biggest mistake. He added to his equations a repulsive term, a "cosmological constant." This anti-gravity force would fill Space/Time and oppose the tendency to collapse. Like epicycles and ether, the cosmic constant would be invisible yet fill all of Space. Einstein provided no theory to explain the constant, and no experiment could isolate it. The constant was a fudge factor added to keep the spherical model from collapsing. The cosmological constant completed Einstein's Static Universe.

Einstein wrote about his spherical Universe in 1917, on the brink of fame. A solar eclipse in May 1919 would provide another opportunity to test General Relativity. British astronomer Arthur Eddington ventured to the island of Principe to observe the eclipse. He returned proclaiming to the press that Einstein's Relativity had been proved. Though the evidence today seems inconclusive, Eddington made the world take notice.

On November 7, a year after the end of World War 1, the London *Times* announced, "REVOLUTION IN SCIENCE." To a world weary from the noise of war, the quiet Einstein became an instant celebrity. In the following year over 100 books appeared about his theory, including Einstein's own *Relativity*.

The rise of his fame began a decline in Einstein's productivity. His great discoveries involving Relativity and the photoelectric effect were behind him. The demands of fame took up much of Einstein's time. Most of his contributions to science would be judging the work of others. In a famous letter to President Roosevelt, Einstein warned of the possibility of an atomic bomb. The job of expanding upon his discoveries fell on other shoulders.

1924 was a very good year for waves. Hawaii's great surfer Duke Kahanamoku won a swimming medal at the Paris Olympics. On the other side of Paris, at the Sorbonne, another Duke named Louis De Broglie made waves with his doctoral thesis. Maxwell had shown that light was an electromagnetic wave. Duke De Broglie suggested that the matter we are made of takes the form of waves, dependent on Planck's value h.

Originality can be bad for an academic career. De Broglie's thesis was so short, simple and important that his professors were ready to reject it. Fortunately De Broglie was an aristocrat, and his older family made the Sorbonne get an outside opinion. Albert Einstein recommended that the Sorbonne give De Broglie his PhD, and no one dared contradict Einstein. 5 years later Einstein would nominate De Broglie for the Nobel Prize—Einstein was a very good friend to have.

Among the eyes who saw the thesis was physicist Erwin Schrödinger. With De Broglie's paper in hand, Schrödinger spent 2 weeks in a Swiss ski resort with a woman friend. After consulting her repeatedly, Schrödinger produced equations for the wave function of matter. The work of De Broglie and Schrödinger showed in detail that particles of matter could also behave as waves, with an energy given by Planck's *h*. Connecting this quantum world with the Universe of Relativity is one of the great problems in physics.

Isaac Newton showed that light waves could be broken up into spectra. Different elements gave off different spectra, and quantum mechanics provided a way of calculating those spectra. The "spiral nebulae" showed spectra that were shifted into the red, as if these objects were receding. Many theories were proposed to account for this redshift, but the answer lay in an expanding Universe.

Alexander Friedmann was born in St. Petersburg, Russia to a musical family. His mother was a pianist, his father a dancer and composer. While studying physics, Alexander became interested in applications to meteorology. During the First World War, he used his knowledge to aid the new tactic of dropping bombs from planes. He flew as an observer on many hazardous missions, becoming known to both sides. On days when Russian bombs were on-target, the Germans would mutter "Friedmann is in the air today."

After the war, despite the distraction of the Russian Revolution, Friedmann devoted himself to the consequences of Relativity. He also believed in the cosmological principle that the Universe looks the same in every direction and that every bit resembles every other bit. Friedmann found solutions to Einstein's equations that indicated an expanding Universe, doing away with the need for a cosmological constant. Sadly Alexander Friedmann passed away in 1925.

Meanwhile a Belgian priest named Georges Lemaitre had both God and the Universe in mind. After attaining the priesthood, he began graduate study at Cambridge University, with Arthur Eddington among his teachers. In a 1927 paper Lemaitre proposed an expanding Universe, with the additional proposal that expansion caused redshift of nebulae. At a conference that year Lemaitre approached Einstein with his idea. The famous scientist could find no flaw with the mathematics, but rejected the idea of expansion. Einstein also pointed out that Friedmann had come up with a similar idea. Several years passed before Einstein accepted that Lemaitre was right.

During 1923-24 Edwin Hubble, working atop Mount Wilson in California, made many observations of "spiral nebulae." His distance measurements proved that they were other galaxies far outside the Milky Way.

There was nothing special about our location in space--our galaxy was just one among many others. Thanks to Hubble, other galaxies became far, far away

At Mount Wilson, Hubble and his colleague Milton Humason observed and catalogued galaxies. They relied upon the light of stars called Cepheid Variables. These stars vary in brightness, with a period related to their luminosity. Observing a Cepheid's period would tell them how much light the star gave off. By measuring how much of that light reached Earth, they could determine distance to the star and galaxy. Cepheid Variables became "standard candles" measuring the distance to their galaxies.

Henrietta Leavitt had previously discovered the relationship between Cepheid periods and their luminosity. Leavitt became interested in astronomy during her senior year in university. She worked as a volunteer assistant at the Harvard Observatory for seven years before being hired for 30 cents per hour. Her supervisors did not want a woman doing theoretical work, so Leavitt examined photographic plates and maintained telescopes. Despite this early challenge, she made many discoveries about the magnitude of stars, including the use of Cepheid Variables as distance markers. Hubble's discovery of an expanding Universe would not have been possible without Henrietta Leavitt.

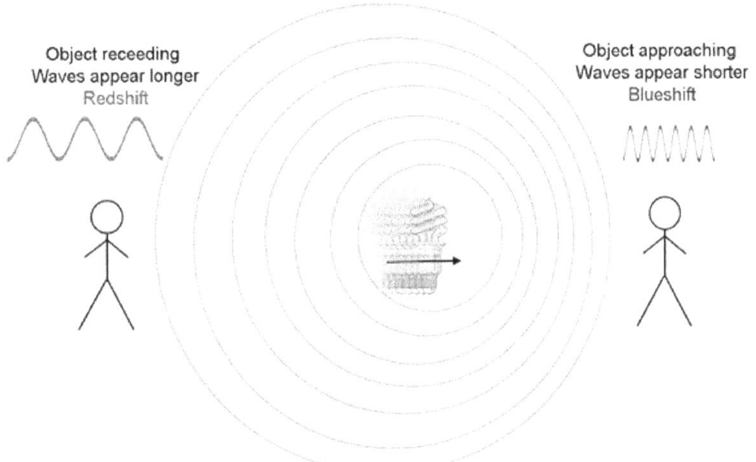

When an object is moving away from us, its light is *redshifted.* This is similar to the Doppler effect, which causes the pitch of a sound to change when the object is moving toward or away from the listener. The light from a receding object is shifted into longer wavelengths, white light would be shifted into red. Redshift is approximately *v/c*, an object's velocity divided by the Speed of Light.

When the redshifts of many galaxies were plotted against their distances, they increased linearly with distance. This did not indicate that our galaxy was unpopular. If the Universe was spherical like a balloon, the galaxies were like spots on its surface. As a balloon expanded, the spots would increase their distance uniformly. A galaxy twice as distant would recede twice as fast. The distance-redshift relation was convincing evidence that our Universe was expanding.

In a 1931 visit to Mount Wilson, followed by an eager press, Einstein conferred with Hubble and peered through the telescope. The light in Hubble's telescope convinced Einstein that the Universe was expanding, forcing him to drop the cosmological constant. Later Einstein would call the constant his "greatest blunder," for it prevented him from predicting an expanding Universe.

I recently consulted with the Einstein statue in Washington. On the tablet in Einstein's hand are his famous equations, minus the cosmological constant. Einstein's statue, never one to blindly follow fashion, has not seen fit to re-introduce the constant. No one today should repeat this constant, for if Einstein called something a blunder he was probably right.

Albert Einstein spent his last decades here at the Institute of Advanced Studies in Princeton. In a final contribution to cosmology, Einstein coauthored to a 1932 paper with Willem de Sitter. In this paper they both favoured a cosmological model that expanded and slowed asymptotically. This Einstein-de Sitter Universe expanded at the rate of its age *t* to the 2/3 power. Mathematically it is the simplest of models. Its expansion continually slows, but will never slow to a stop or reverse. Density of this Universe is a "critical" value called Omega or Ω, neither too light nor too dense. Einstein and De Sitter also suggested that large amounts of mass remain undetected. Einstein-de Sitter was long the cosmologist's favourite model.

Einstein's early work was prodigious, but all things slow with time. Einstein spent his twilight years as the world's most famous scientist, but personally frustrated. He spent years trying to find a Unified Field Theory to explain the Universe simply. He did leave us one final means of judging ideas. When asked what he would think if a theory of the Universe was complicated, Einstein replied, "then I would not be interested in it."

In 1948 George Gamow made one of the most important predictions of cosmology. If the Universe had once been smaller and hotter, it would have been dominated by radiation. Gamow predicted that some of the radiation should still be around, coming from every point in the sky. Gamow greatly overestimated the temperature at 50 degrees Kelvin. If the temperature were known, Planck's blackbody formula could predict the spectrum of that radiation. Because Gamow's work was largely ignored, the two physicists who discovered this radiation had no idea what they had found.

This photo is of Echo, one of the first communications satellites. It was a sphere made of aluminum-coated mylar, like a party balloon. It was launched into Space and inflated so that experimenters with microwave antennas could bounce signals off Echo. For a balloon supported by pressure, a sphere is also the most natural shape.

The Holmdel Horn Antenna in New Jersey was built in 1959 for bouncing signals off the Echo satellite. In 1965 Arnold Penzias and Joseph Wilson from Bell Laboratories were calibrating the antenna. As with all radios, there was a background of hisses and pops to be removed. The two physicists could not account for a signal which came from every direction in the sky. The temperature of this radiation was 3 degrees Kelvin, close to predictions of expanding Universe models. Discovery of the Cosmic Microwave Background was "smoking gun" evidence that the Universe grew from a hotter, denser state.

Aristotle would marvel at the amount of data scientists collected. As evidence mounted in favour of the expanding Universe, there was still no guiding principle. The master Plato would ask, why does the Universe expand? An answer could tell us both how it began and whether it will end.

A Simple Principle

In imagining the spherical Universe, Einstein was far ahead of his contemporaries, as far as was possible given the observations of his time. Today we have far more evidence to work with. The Universe appears to have a beginning, commonly called a "Big Bang." Near this time of great heat and density, an enormous mass was confined in a very small Universe.

This beginning can be drawn as a point. Cosmologists agree that this baby Universe had a finite size. Therefore it had a finite mass **M**.

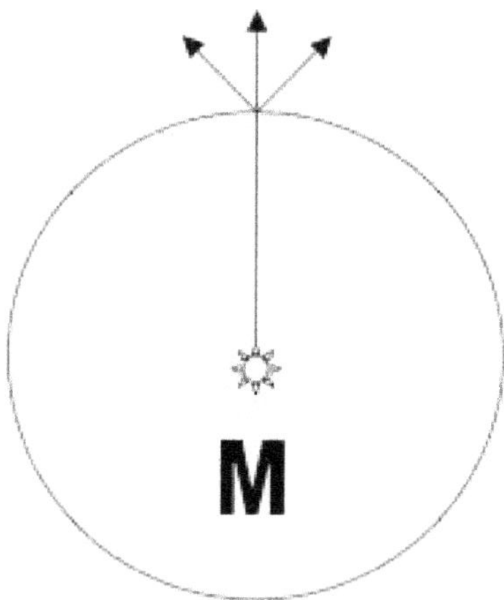

Near the Big Bang, all points of this Universe were very close to one another. Our separation from the Big Bang is a matter of Time, not Space. Astronomers have estimated that age t is about 13.8 billion years.

Our light cone represents the local conditions of Special Relativity. Our mass, even the mass of our galaxy, is negligible compared to **M** of the Universe. We are within the light cone of that enormous mass, and its gravity affects us. As Newton showed, the gravitational force from a spherical mass distribution is the same as if all that mass were concentrated in a point.

This drawing, simple as a child's thought, represents the entire Universe. Our three dimensions x, y and z are again compressed into the page. Any direction that we can travel in Space keeps us in that narrow circle. The Universe might appear infinite and flat to our experience, but may still be curved in the fourth dimension.

From a sphere, the combined gravitational attraction is the same as if everything was in a central point. There is no centre in Space, for every bit resembles every other bit. There is a centre in Time, which we call a "Big Bang." As time t increases, our Universe expands away from the Big Bang.

Newton's Law of Gravitation insists that mass affects us at a distance. It is meaningless to express this attraction over Time, so we must add the factor c. Here we expand on one Principle of Special Relativity. Since Space and Time are one phenomenon related by c: Scale R of the Universe is its age t multiplied by c.

Near the Big Bang the Universe was tiny--all points within it were from our point of view the same distance away. Our separation from the Big Bang's centre is simply the age *t* of our Universe. Expansion of the Universe is then indistinguishable from the forward flow of Time. This simple principle tells us why we live in an expanding Universe: As time *t* increases, scale *R* expands.

A simple principle also explains an "Arrow of Time." Most laws of physics are time-symnetrical; they apply equally forward or backward in Time. Expansion does not share this symmetry. As Time increases, the Universe only expands outward. Scientists call this the Cosmological Arrow.

Since that distant time of the Big Bang, our Universe has been expanding. It can't expand at the same rate continuously, for gravity slows it down. Science would not be complete if it did not make predictions.

The Speed of Light *c* must be further related to time *t*.

This leads to one equation with many predictions.

NEXT: The Only Equation in the Book

6

The Only Equation In the Book

When forced to summarize the general theory of relativity in one sentence, **Time** *and* **Space** *and* **Gravitation** *have no separate existence from* **Matter**.—Albert Einstein

Mathematics can be both beautiful and puzzling. A truly beautiful equation can describe many phenomena with a single line. Equations can also frighten people completely away from science. Many budding scientists and engineers have abandoned those dreams after glimpsing the amount of math involved. The mathematics of epicycles was incomprehensible except to a learned elite. Many layers of complexity kept the theory of invisible spheres unchallenged for centuries. The most lasting equations are simple.

This book contains just one equation. One authority claims that every equation in a book cuts the number of readers in half. It would be tempting to write a book with no equations at all. Such a book could make grand promises about theories of everything, without offering anything that could be proved by experiment. While grand promises can be good for careers, the purpose of science is to describe the world in simple rules.

Since there is only one equation in this book, it should describe as much as possible. It should be simple enough for the young people reading this book to grasp. Including an equation from anyone but this author would be quite improper. I will omit the many years of calculations to reach this equation. It is pictured with a spherical Universe similar to that of Pascal, Poe and Einstein.

"*When forced to summarise the General Theory of Relativity in one sentence, Time and Space and Gravity have no separate existence from Matter,*" Albert Einstein said. Today we can translate that statement into an equation:

$$GM=tc^3$$

Where G is Newton's gravitational constant, M is mass of the Universe, t is its age and c is the Speed of Light.

The drawing and equation fit on a T-shirt, yet tell us everything we wanted to know about the Universe but were afraid to ask: How big it is, how fast it expands, and whether it will ever slow to a stop. This allows us to make predictions about the density of mass, the proportions of different kinds of mass, the redshifts of distant supernovae, and many other phenomena. It is pleasing when our big and mysterious Universe can be described simply.

We can describe the equation bit by bit. Newton's gravitational constant G has so far appeared to be constant. Arthur Eddington and Paul Dirac both suggested hypotheses of varying G. These theories were disproved by many experiments, including the first spacecraft to land on another planet. Data from the Viking Lander showed that Mars' orbital period was not changing, as it would if G varied. The best evidence for a constant G is the existence of life on Earth. If Newton's G changed with time, Earth's orbit and temperature would not be stable enough for life to evolve.

Since this Universe has a finite size, the mass M contained within is immense but also finite. This quantity M may change slightly, for small amounts of mass are created or destroyed in nuclear reactions. During the hot times near the Big Bang, this may have happened quite often. In its present cool state, the Universe has a nearly constant mass.

The Universe is big—but we can envision how big from pure math. Since G and c can be measured in a lab, and age t can be estimated from astronomical observations, we can figure out how massive everything in the Universe is. That mass M is over 100 billion times the mass of our Milky Way galaxy. Since G and M are nearly constant, both sides of the "equals" sign are also constant.

The dimension of Time is represented by t. Once people thought that this age was only a few thousand years. Only in fairly recent history have we known Earth and the Universe to be billions of years old. Einstein at first thought the Universe was eternal and unchanging, leading to his blunder of a cosmological constant. Today evidence from redshifts to the microwave background all indicate that the Universe has a finite age. At this writing, the best estimate of age t is about 13.8 billion years.

Three dimensions of Space are represented by c, the Speed of Light. As we have seen, c is a *conversion* factor between Space/Time. Since both sides of the 'equals' sign are constant, as t increases c slows down. When t was tiny c was enormous, and the Universe expanded like a "Big Bang." As t increased, that expansion slowed due to gravitation, and continues slowing to this day. Now that t is billions of years, change in c is very small. Though today's Speed of Light is 299,792 kilometres per second, in one year it would change by only *0.72 centimeters per second*! A simple equation allows us to predict the value of c in the past, present, or future.

Expanding Space/Time

First we can solve for the scale R as a function of time. The expansion rate turns out to be proportional to $t^{2/3}$, as Einstein and de Sitter believed. The Universe expands and slows asymptotically; it will never stop, reverse or collapse. This tells us that there is nothing particularly special about the time we live in. The Universe has been behaving according to $GM = tc^3$ since it was no bigger than the drawing, and will do so indefinitely.

As Einstein believed, the Universe follows the Perfect Cosmological Principle. Every bit of the Universe resembles every other bit. Not only does it follow the Principle in three dimensions, but in the fourth dimension of Time. The Universe looks more or less the same at any time. Surprisingly, the "Big Bang" has not ended. We are still in it, part of a continuing process of expansion.

When scale R is plugged into Alexander Friedmann's equations, the "critical" density Ω of Einstein and De Sitter pops out. This is not just a critical density but the most stable density. If the Universe were below this density, quantum theory predicts that matter would form via pair production. In this process, particles and antiparticles would form until the stable density Ω was reached. The Universe has exactly the density to expand and slow.

The stable density was not the initial density. Before any matter formed, the Universe was underweight. The matter we are made of—protons, electrons, and neutrons, would have formed to fill this gap. The difference, between density before matter and density with matter, can be calculated from pure math. That proportion is:

The matter that we are made of is that small slice of the Universe.

The First Law of Thermodynamics states that energy is conserved—it cannot be created or removed, just changed into different forms. Every object before our eyes contains energy. Newton showed that objects have both potential and kinetic energy. Einstein added a rest energy given by his famous equation.

When all these energies are added up, the total is Zero! This result applies to any object, from the tiniest particle to the heaviest galaxy. Even photons, the massless particles of light, have the same total energy. Our enormous, complex Universe has zero total energy. It is the ultimate free lunch, which has allowed it to expand from a tiny point to an enormous size.

The Cosmic Horizon

Humans may have first suspected that Earth was spherical because of objects disappearing over the horizon. When the Speed of Light is known, the distance our eyes can see can also be calculated. Scientists call this the cosmic horizon, the distance light has traveled since the Universe began. This turns out to be exactly (3/2) times the radius. On Einstein's sphere that is almost halfway around. If our position were the North Pole, we could see nearly to the Equator. Light that reaches us from near the Equator dates from a time near the Big Bang. The cosmic horizon

expands at the same rate as the rest of Space/Time.

"The Dark side of the Moon" is a title for a Pink Floyd song, but the Moon's far side isn't really dark. It receives as much sunlight as the near side, but was hidden from human eyes until we could send spacecraft to observe it. Like the Moon, the Universe has both a visible side and a hidden side that is beyond our sight. The Moon offers some of the best evidence that light is slowing down—more about that in Chapter 8.

If the Speed of Light *c* did not change, the visible patch of the Universe would grow bigger. Astronomers would see distant objects suddenly appear in the sky! After a long enough time, our telescopes would see completely around the sphere to view ourselves. Such a Universe could not support life, for the radiation from its beginning would make its way around the sphere to cook us.

c and h

Nearly all things change over time. If the Speed of Light *c* varies, something else must change. Planck's value *h* is critical to quantum mechanics--it allowed Max Planck to explain the blackbody spectrum, and Einstein to explain the photoelectric effect. It is also a key to the Uncertainty Principle, which says that particle positions cannot be measured exactly.

Speed of Light *c* and *h* are multiplied together in the fine-structure constant α. This constant is important in predicting spectral lines from stars like the Sun. *c* and *h* also appear together in Chandrasekhar's Limit, which relates to exploding stars. Because our Sun's mass is below this limit, it will never explode into a supernova. As Einstein found, *c* and *h* appear together in the energy of light particles.

Most measurements indicate that the product *hc* is indeed constant. This is not an obstacle, but a first link between Relativity and Quantum Mechanics. If *c* is slowing but the product *hc* is constant, then Planck's value *h* must also increase. The quantity *h* is in fact the solution to a *random-walk* problem in three dimensions, something physicists use for subatomic particles.

Measuring change in *h* would be even more difficult than for the Speed of Light, for Planck's value is only revealed in the microscopic world. Ole Roemer measured *c* back in 1676, but not until 1900 did Planck find *h*. Many undergraduates struggle to remember the Planck value. Fortunately, most students know the present value *c*. It is an easy step to remember *hc*, which is just 2.0×10^{-25}.
If a student divides by *c*, she will get Planck's value *h* correct to two decimal places.

Planck's value *h* is a key to the other arrow of time, the thermodynamic arrow. Nature is full of processes that are irreversible. If a plate is dropped from a high place, it shatters into many pieces but never reassembles itself. Heat from a fire spreads throughout a room but does not converge again at the fire. These are demonstrations of entropy, the tendency for disorder to increase. These thermodynamic processes are governed by the value *h*. An increasing *h* means that the entropy of the Universe increases with time. Its complexity, and the creation of structures that lead to life, also increases.

In the Beginning: Initial Conditions

From childhood we wonder where the Universe came from. This question is the basis of cosmology and many other sciences. Every culture has created legends to try and explain how the Universe was created. Today we can imagine some initial conditions for how the Universe began. These initial conditions led to the complexity we observe today.

In the beginning the Universe was indeed without shape or form. There was no Space or even Time. From the perspective of this beginning, every direction was forward in Time, like looking outward from a single point. Anything before this is beyond human imagination, for not even Time itself existed. The Universe began at a time of Zero.

At time Zero the Universe also had zero radius, like a point. The Planck value h was also zero, for with zero scale there is no uncertainty in position of anything. Creation began with the Speed of Light. When time t was near zero, c was nearly infinite. This immense speed caused the Universe to expand like a Bang. As today, the total energy of the Universe was zero. There was no need for any additional energies-these initial conditions were all that was needed to cause the Big Bang.

Collapse and Creation

The Universe could not explode at such an immense rate for long--gravity almost immediately started to slow its growth. The Planck value *h* also grew, though not as fast as the Universe. Though the overall energy was Zero, quantum uncertainties ensured that the Universe did not have the same density everywhere. These fluctuations were magnified by the rapid expansion.

Some of these density fluctuations grew so large that they collapsed due to their own gravity. Gravity from these Black Holes sucked in the mass around them, clearing great voids in Space. Almost nothing could exist in these voids except for the huge masses in their centres. As the Universe expanded, some of these voids grew to immense size.

Primordial Black Holes formed in a variety of sizes; seeding formation of voids, clusters, galaxies and even smaller objects. Billions upon billions of tiny Black Holes were also created. Formed shortly after the Big Bang, they have never been matter, and would be nearly invisible to our eyes. Drawn by gravity toward larger objects, they would form dark haloes around galaxies.

The matter that we are made of formed next. After the first Black Holes had formed the Universe was still underweight by *4.507034%*. The Universe was very dense and hot with radiation from billions of Black Holes. Quantum mechanics predicts that this Space was filled with virtual particles winking in and out of existence.

Enough of these virtual particles remained to bring the Universe up to its stable density. This matter would form stars, planets and eventually life. The first atoms formed within minutes of the Universe's beginning--the details are beyond the scope of a small book. The wonderful complexity of matter we create every day.

The language of mathematics can be quite powerful. A single equation may describe an entire Universe. The most beautiful equations would be useless to our understanding if they did not make testable predictions. Science is just speculation if it is not supported by results. Extraordinary claims require extraordinary evidence. Next we will see the results of many experiments.

NEXT: The Proof In the Pudding

7

The Proof In the Pudding

Darkness cannot drive out darkness,
only light can do that--Martin Luther King

Supernova image courtesy NASA

The Universe has been compared to raisin pudding, where the raisins are galaxies. As the pudding expands, the individual raisins grow further apart. A raisin twice as far away recedes twice as fast. The more distant an object, the faster it recedes from us, causing its light to be redshifted. We are somewhere within that pudding, looking for clues.

Like the Renaissance or the beginning of the 20th century, today is an exciting time to be in science. Spacecraft have extended humanity's reach into the solar system. New discoveries and data arrive almost daily. Like a baby's first months, observations come so fast that scientists don't know what they are seeing. Hopefully this book can convey the thrill of discovery.

In case the reader has skipped the last chapter, The Only Equation in the Book predicts that light is slowing down at a very tiny rate. The duty of scientists is to make testable predictions so that theories can be adopted or discarded. Change in *c* is not a new idea, dating to at least 1874 and Lord Kelvin. In recent years, a growing number of physicists have questioned whether the Speed of Light has always been the same.

Here are described data from many spacecraft and experiments. Several lines of evidence support the prediction of a changing Speed of Light. In each case, alternate explanations for the data are mentioned. For science to advance, ideas deserve a fair hearing. A first picture comes from shortly after the Big Bang.

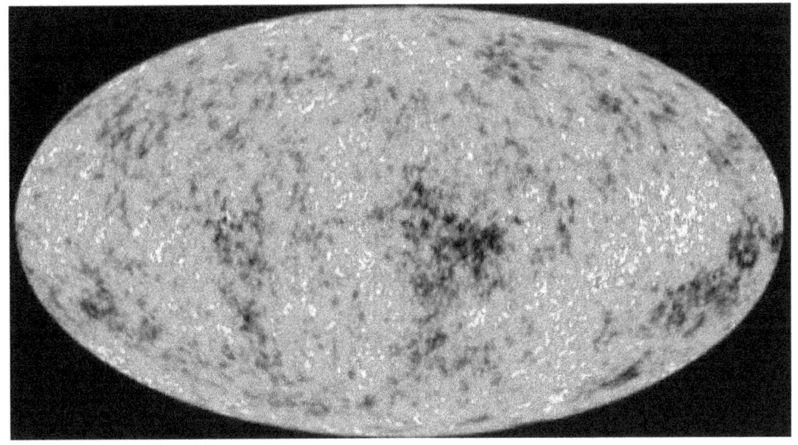

Courtesy WMAP

The Cosmic Microwave Background

This is light from near the Big Bang, predicted by George Gamow then found by Penzias and Wilson. Our picture of the Cosmic Microwave Background comes from the Wilkinson Microwave Anisotropy Probe (WMAP). This spacecraft orbited at the L2 point, on the other side of the Moon. Its mission was to make detailed temperature maps of the background sky.

Penzias and Wilson's accidental discovery was radiation from 380,000 years after the Big Bang. Expansion and Time have stretched its wavelength and cooled its temperature to just 2.7 degrees above absolute zero. The COBE spacecraft and BOOMERANG balloon experiment made previous maps of this background. Recently the European-built PLANCK spacecraft has mapped the CMB in even more detail.

The most striking feature of the microwave background is its uniformity--to one part in ten thousand. Shades in the picture represent very tiny fluctuations. If we look at this picture, the average temperature is nearly the same over large parts of the sky. This indicates that large regions of the Universe were within the range of light signals. Even 380,000 years after the Big Bang, the Speed of Light was much larger.

In the late 1970's, when rising prices were on everyone's mind, "inflation" of the Universe was proposed. Physicists assumed that the Speed of Light was always the same, but when the Universe was just 10^{-33} seconds old it suddenly expanded at warp speed, much faster than light. This idea has been a useful step, but it is not a theory. "Inflation" would violate both the First Law of Thermodynamics, which states that energy is conserved, and Relativity's stipulation that nothing travels faster than light.

"Inflation" relies on imaginary repulsive energies to break these laws. These "inflatons" or would have appeared at the Big Bang, yet conveniently disappeared before we could observe them. While growth of prices was tamed, the inflated idea has taken over. It has spawned an inflation of different theories, but a mechanism is still not known. Scientists cannot time-travel to the first 10^{-33} seconds to observe inflation. No conceivable experiment can approach the titanic energies near the Big Bang. Since nothing like "inflation" has ever been observed in nature, it cannot and will never be proven.

The "inflation" idea claims the infant Universe abruptly expanded to 10^{43} times its original size! While it began tiny and spherical, it expanded so large that it became flat. Quantum fluctuations in density would expand to almost infinite size. Using the equations of spherical harmonics, theorists can predict the spectrum of these fluctuations.

Results from two satellite experiments, WMAP and COBE both showed a spectrum far different from inflation's prediction. Their data falls outside the error bars. Most striking, the graph is nearly zero for angles greater than 60 degrees. Inflation with a flat universe predicts that they should extend to infinitely high angles. As a ship disappearing over the horizon shows that Earth is spherical, the cutoff indicates that the Universe is also curved. Large objects, even raindrops, tend to form spheres. The Universe is very large.

The principle from Chapter 5 states that Scale R of the Universe is its age multiplied by c. Any changes in density would also be of limited size. When Theory's prediction is plotted on the graph, it fits the data precisely. The close fit indicates that the Universe is curved with radius R, as predicted. When prediction fits the data this closely, it may mean something! WMAP scientists conclude *"An alternative model would be an exciting development."*

Light from the cosmic background provides a huge amount of information. Peaks in the graph indicate the overall density of the Universe, and even proportions of different kinds of mass. A first peak indicates the overall density. That turns out to be exactly the "critical" density for Einstein-de Sitter expansion, which is also the stable density predicted by Theory. Succeeding peaks indicate the amount of density that is the matter we are made of. This is predicted by Theory to be *4.507034%;* WMAP's result is *4.4 ± 0.3%.* The amount of mass that would be invisible, having been sucked into giant Black Holes, can be predicted at about *68.3%.* This exactly fits data from the CMB.

The huge amount of microwave data allows many interpretations. It can be fashionable to say that the Universe is flat, like the Earth. Flatness can appear to be supported by data. A typical distance between peaks is about one degree. The distance light has traveled since the Big Bang was assumed to simply be *ct*, age of the Universe multiplied by the present Speed of Light. Combined with the distance between peaks, this gives three sides of a triangle. The angles total 180 degrees, seeming to prove that this triangle was on a flat space.

The triangle example, seemingly solid as Pythagoras, does not consider that the Speed of Light changes. The distance light has traveled is not *ct,* but *(3/2)ct* . The angles do not add up to 180 and the triangle is not flat. The greater distance indicates that the Space is curved, and how much.

Again we take two peaks separated by one degree. Instead of on a flat map they are near the equator of a globe with radius ct. Our location would then be on the North Pole, nearly 90 degrees away. If we draw curved lines out to those two peaks on the equator, they will appear one degree apart. They will only appear one degree apart of the Universe is of radius *ct.* Including the change in *c*, the data shows that the Universe is curved with this radius. It is not mathematically possible for the Universe to expand from a tiny point into a flat plane. Such a universe would also have infinite mass, and could not expand at all.

In November 2019, shortly after I gave a talk in Paris, researchers led by Eleonora de Valentino of University of Manchester and Joseph Silk of Oxford University announced a study concluding that the Universe is curved. "Inflation" was thought to be untestable, but every single inflated idea predicts that our world is flat. A curved Universe disproves "inflation".

Unlike "inflation" a proper theory makes a testable prediction that the Speed of Light is slowing. The microwave sky is strong evidence for change in *c*. Data from this background can be interpreted in more than one way. Inflation with flat space can be fitted to observations, but is ruled out by the scale of density fluctuations. A spherical Space/Time makes an even more precise fit to data, and also predicts the densities of mass. Signs of *c* change come from huge objects not quite as distant.

Galaxies

Every galaxy, like our Milky Way, contains at its centre a massive Black Hole. Giant Black Holes very likely exist in Space without surrounding galaxies. Instruments like the Hubble Space Telescope and the Subaru Telescope atop Mauna Kea can find galaxies that existed only a few hundred million years after the Big Bang. They are so distant in Space/Time that their light has shifted into the infrared. In the distant past we have seen quasars and Active Galactic Nuclei, powerful objects each radiating more light than a million galaxies. These primordial objects are powered by massive Black Holes.

Near the beginning of the Universe, not enough time had passed for Black Holes to evolve from stars. Primordial Black Holes could have formed by collapse of quantum fluctuations. In the immense densities near the Big Bang, fluctuations were big enough to form Black Holes without becoming matter first. Many scientists, like the late Stephen Hawking, believe that Primordial Black Holes survive today.

Scientists did not consider the Speed of Light. The mass of a collapsing Black Hole is limited by a cosmic horizon, the distance light can travel in a given time. Previously it was thought that Primordial Black Holes would be relatively tiny, because of light's limited reach. If we consider that the primordial Speed of Light was faster, the mass within the horizon was enormous. Primordial Black Holes could have formed of immense size, seeding formation of galaxies. The massive Black Hole at the centre of our Milky Way may be older than the galaxy.

Though we live in the Milky Way and see its band of stars at night, astronomers do not understand how our galaxy formed. It had been thought that the Milky Way formed from the merger of smaller galaxies. An international team led by astronomer Manuela Zoccalli has disproved that idea. Her observations show that stars in the galaxy's central bulge have a different chemical composition than stars in the spiral arms. If the Milky Way had formed from collisions of smaller galaxies, stars of similar ingredients would be mixed together everywhere we look. Stars from the outer arms would have been pulled toward the centre. Zoccalli's observations also show that our galaxy's central bulge formed in an amazingly short time, less than a billion years.

Observations from the Subaru Telescope atop Mauna Kea have found galaxies formed barely 500 million years after the Big Bang. As we look farther into the past, we continue to see signs of massive Primordial Black Holes. These objects could not form without a higher value of *c*. If these objects date from near the Big Bang, they are evidence that *c* was once much higher. If *c* had not been higher, our Milky Way galaxy could not have formed. More precise and controversial evidence comes from explosions in galaxies.

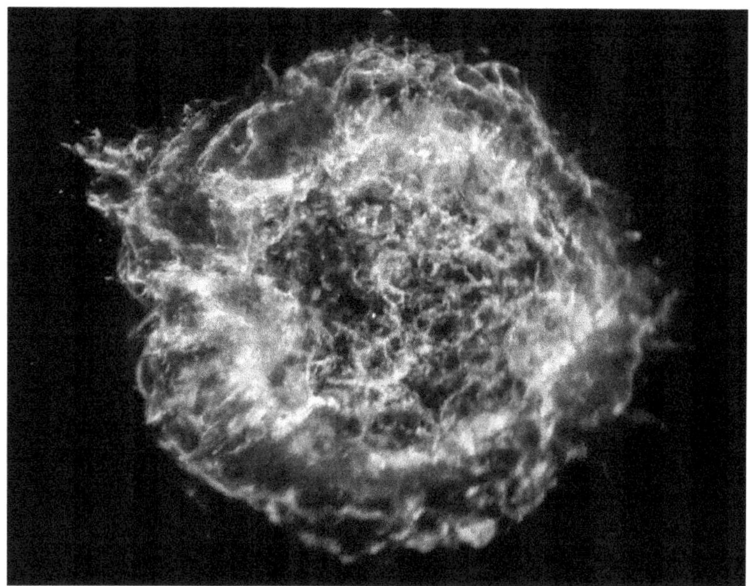

Supernova

When a star explodes, its light can briefly outshine an entire galaxy. When Tycho Brahe saw a supernova in 1572 he could only speculate what it was. Though the power source of these huge explosions has not been understood, enough is known from observations to make predictions about their behaviour. Observers who find exploding stars stumble to interpret what they see.

Supernovae have been classified into various types. Physicists have focused on Type Ia supernovae, which have been observed in many galaxies and are thought to all have the same luminosity. From observations, predictions have been made as to their spectra and brightness. Though their power source was not understood, Type Ia supernovae were considered as "standard candles".

Edwin Hubble, building on the work of Henrietta Leavitt, used Cepheid Variable stars to show that galaxies were receding. By searching for supernovae at high redshift, observers hoped to extend the Hubble diagram further into the past. Redshifts of nearby supernovae increased in a straight with distance, indicating that the Universe expands. When observations were extended to high redshifts, the line curved mysteriously upward, as if the Universe were accelerating.

The lead authors of two "independent" groups were both working from the same campus, so they could check their findings. The lead author of one group's paper was in Campbell Hall at University of California, the other group's lead author was on the same campus at Lawrence Berkeley Laboratory. At least one man belonged to both teams. It benefited them claiming to be two teams.

Two large observing teams did not consider whether Speed of Light *c* had affected their measurements. They concluded that velocity *v* and the Universe was accelerating. Because acceleration would violate the First Law of Thermodynamics (Conservation of Energy) Einstein's "blunder" of a cosmological constant was revived. Since data showed that the cosmological constant could not be constant, it was renamed as "dark" energy.

Like epicycles and ether, this energy would be invisible yet fill most of Space. Some researchers dug up Aristotle's millennia-old idea of a "quintessence". Unlike all other forms of mass and energy, which attract each other gravitationally, "dark" energies would exert a repulsive force. No experiment could isolate this invisible actor; there is not a single sample or track in a bubble chamber to prove its existence. "Dark" energy is often called a fudge factor.

Redshifts of supernovae are the only evidence of "cosmic acceleration." The microwave background, dating from a time only 380,000 years after the Big Bang, tells us nothing about acceleration. The supernova observers did not know what they were seeing. (At this writing, they've still not figured it out.) Though their observations were highly valuable, two large groups of scientists did not consider what a child could ask: What if *v/c* increases not because *v* accelerates but because *c* slows down?

Near the Einstein statue in DC is the monument to another Nobel Prize winner, Dr. Martin Luther King.

On the marble wall is a quote from Dr. King:

"Darkness cannot drive out darkness,
only light can do that."

The answer to "acceleration" may not be in imaginary darkness, but in the Speed of Light.

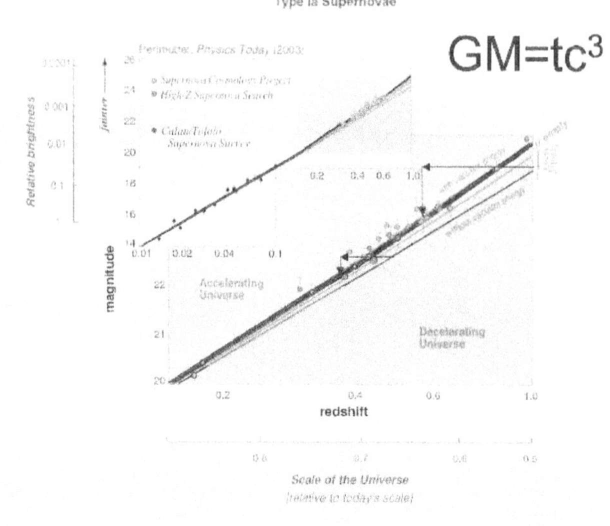

Courtesy of Supernova Cosmology Project

An object of redshift *1.0* recedes at *60%* of today's Speed of Light. That would be only *42%* of *c* at the time that light was emitted. The redshift we observe will be only *0.57.* (Horizontal arrow) Supernovae convert mass *m* into energy *E* by the famous Einstein equation with mc^2. For a supernova of redshift 1, that energy would be doubled, changing its magnitude by -*.75.* (Vertical arrow) The prediction line is hard to see, because it is covered with data points. Given the observer's error bars, the prediction intercepts *95%* of the points. The "acceleration" curve can be precisely predicted from The Speed of Light, without imaginary energies. Every point on the graph indicates that light has been slowing down.

Professor Subir Sarker of Oxford University has never believed in "dark" energy. In October 2019 as I delivered my talk at Institute d'Astrophysique in Paris, Sarkar and IAP scientists published a study concluding that "dark" energy doesn't exist. In January 2020, as I attended the American Astronomical Society meeting in Honolulu, Young-Wook Lee of Hansei University in Korea and his colleagues announced another study showing that cosmic acceleration is an illusion. A growing number of scientists doubt the existence of "dark" energy.

It is still possible to profess that the Universe accelerates because of some repulsive "dark" energy. If scientists turn to that dark force, they may follow its dark path for years or decades. A complicated equation of state for "dark" energy can also be produced to match the data. Just as enough epicycles can put Earth at the centre of the cosmos, they can make the Universe appear to accelerate.

Speculations of "inflation" and "dark" energy lead not to a solution but to a divergence of theories. Both mysteries may be explained by a slowing Speed of Light. To verify that c has changed over time, corroborating evidence is required from truly independent sources. That may come from the nearest star of all, and from our Moon. The proof is in the pudding.

This is the view someone would have seen July 22, 2010, 5:51 PM Universal Time, just before local sunrise at Gusev Crater on Mars. On that day Earth appeared to rise before the Sun, leading a parade of planets followed by distant Jupiter (centre) and Venus. If Earth's Moon were bright enough to be seen, its orbit would span just 1/6 the diameter of the Moon seen from Earth. All of human history, from life's beginnings to the first steps into Space, took place on that tiny light in the sky.

A mind on Earth, a single light on that small planet, can understand the size of the Universe.

Next: Sunlight and Moonlight

8

Sunlight and Moonlight

A single candle can both define and defy the darkness.
—Anne Frank

Sunrise on Mount Haleakala, Maui

The Sun greets us each morning. "Sunrise" is a misnomer; the Sun appears to rise because of Earth's rotation. Though today we know that Earth orbits the Sun, the term has stuck. Our eyes see the world thanks to light from the Sun. Before the invention of fire, the Sun and Moon were our only sources of light. Heat from the Sun has allowed life to evolve on Earth's surface. The Sun and its life-sustaining heat also tell us about the Speed of Light.

Scientist George Washington Carver, who was also compared to Leonardo da Vinci, enjoyed the light of sunrise. He rose every day before dawn to walk through the woods. The plants and crops grown in the Sun's light fascinated him. Life on our planet's surface is deeply dependent upon the Sun.

According to theories of astrophysics, life should not have evolved on Earth at all. Physicists' models of the Sun say that several billion years past the Sun shone with only *75%* of its present luminosity. Earth's average temperature would have been *-15 degrees* Celsius, frozen solid. An ice cube planet would reflect sunlight into Space, causing even colder temperatures. Evolution of life would have been very unlikely.

When we look up from the equations and take a walk before dawn, the variety of life seems extraordinary. Study of the Earth shows sedimentary rocks, produced by liquid water, nearly as old as Earth itself. Other geologic clues confirm the presence of rivers and seas 4 billion years ago. Paleontology dates the earliest life forms on Earth at least 3.4 billion and possibly 4 billion years ago. Clearly liquid water and life both existed when the Standard Solar Model says Earth was colder than the movie *Frozen*.

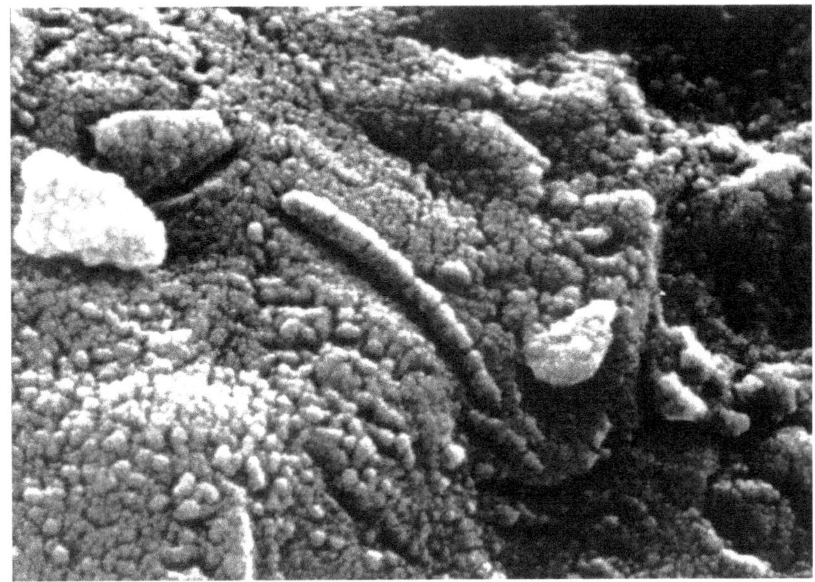

3.5 billion years ago, while life was gaining a foothold on Earth, rocks on Mars melted and crystallized. 15 million years ago a meteorite impact caused one small rock to be blown free of Mars and wander through the solar system for eons. 13,000 years ago the rock struck Earth in the Allen Hills of Antarctica. It remained in the ice until found by NASA Geologist Robbie Score in 1984. For years meteorite ALH84001 sat on a shelf until geologists realized that it came from Mars.

I had the honour of working with and being close friends with the late Dr. David McKay of NASA, who passed away in 2013. In addition to training me, Dave McKay trained the first astronauts to walk on the Moon. In 1996 a team led by Dr. McKay found within ALH84001 indications of fossilized life. This was the most striking evidence yet found that life on Earth was not alone.

In 2005 geologist Vicky Hamilton at University of Hawaii concluded that the rock was from Valles Marineris, the immense Martian canyon over 5 miles deep. This meteorite indicates that early Mars also had conditions suitable for liquid water and life. Since the McKay team's discovery about ALH84001, photos from spacecraft have shown signs of ancient streams, river valleys, and even oceans on Mars.

Standard Models of physics can be incorrect. Both Earth and Mars show signs of water and life when the Standard Solar Model says they were *Frozen*. The fact that life exists to read this book conflicts with physicists' model. This conflict is called the Faint Young Sun Paradox.

At one time scientists thought that huge amounts of carbon dioxide in Earth's early atmosphere raised the temperature. This greenhouse effect would have to be precise as a thermostat to keep our climate comfortable. This custom heating, like "dark" energy, was only an inference. There is no direct evidence for that much CO2 in Earth's atmosphere.

Geologists Robert Rye, Phillip Kuo, and Heinrich Holland have studied the amount of carbon in early Earth's soil. They concluded that the maximum amount of CO2 that could have been in Earth's atmosphere was far below the amount needed to offset the "Faint Young Sun." Other gases, such as methane, pose similar problems. CO2 is unlikely to have custom-heated two planets for life.

It should be disturbing when astrophysics predicts that life can't have evolved. Fortunately, the Speed of Light can help save the Standard Solar Model. The Sun also turns its fuel into energy according to the famous Einstein equation with mc^2. If the c were faster when Earth was formed, both the Sun's luminosity and Earth's temperature would have been higher than originally thought.

Presently Earth is estimated to be 4.6 billion years and the Universe 13.8 billion years old, about 1.5 times its age at time of Earth's formation. Instead of just *75%,* the Sun shone with *99%* of its present luminosity. The "solar constant may indeed be constant, allowing our life to have evolved on Earth for billions of years.

If life survived on Mars today, it would be hidden underground. Earth is fortunate to be protected by a magnetic field from the radiation of Space. So much radiation fills the Universe that it poses a health hazard for astronauts. James Clerk Maxwell showed that light is one form of this radiation. Despite the apparent dark of Space, we live in a Universe of light.

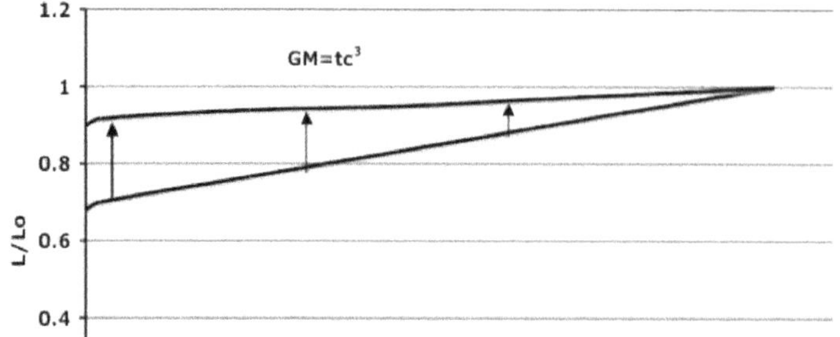

Earth's thermometer measures the Speed of Light. On this graph, the Standard Solar Model of luminosity is an upward curve starting near *75%* and rising to reach today's level. When *c* change is included, the luminosity curve becomes nearly level, centred within our comfort zone. Instead of *75%,* solar luminosity was almost exactly what we enjoy today. (For cosmologists, Earth's history covers redshifts of *0.3* to *0*.) This graph precisely corroborates supernova evidence, distinguishing $GM = tc^3$ from "accelerating universe" ideas and models where *c* is higher only near the Big Bang. If *c* had not changed, we would not be here to talk about it. Sun and supernovae provide two corroborating sets of data from entirely different sources, confirming that *c* has changed in the amounts predicted.

This should be of great comfort in the future, for the Standard Solar Model also predicts that the Sun will continue to get hotter until Earth's oceans boil. Old models claimed that we occupy a lucky time when Earth is neither freezing nor boiling, just as they put Earth in the centre of the Universe. Thanks to the Speed of Light, Earth's temperature will be comfortable for a very long time. Our presence on Earth would not be possible if light had not slowed down.

Exploding stars and our Sun tell us that light has slowed. We have more ways of measuring change in the Speed of Light. Galileo tried timing light with lanterns on hilltops, but lacked a distant enough hilltop or a good clock. A distant hill has been available since July 20,1969.

Apollo 11 Lunar Laser Ranging Experiment

The Moon

One of the great achievements in human history climaxed when Neil Armstrong and Buzz Aldrin set foot on the Moon. Dr. David McKay trained both men, and was the only scientist in NASA Mission Control when Armstrong walked on the surface. On the 50th anniversary of Apollo 11, Neil and Buzz visited us at Johnson Space Center. While studying the Moon as a Scientist, I published calculation of a lunar anomaly, verifying change in the Speed of Light.

The Lunar Laser Ranging Experiment left behind by Apollo astronauts is composed of quartz corner reflectors like those on a bicycle. By bouncing laser beams from Earth observatories, astronomers can measure the Moon's distance with great accuracy. Data from the LLRE has told us that the Moon still has a liquid core, verified that Newton's G is indeed constant, and provided one more test of Einstein's General Relativity. LLRE may also have stumbled onto a discovery about light.

Nearly all things, even the Speed of Light, change with time. The Moon's distance from Earth, about *384,402 km*, has long been known to be increasing due to tidal forces. As the Moon raises ocean tides on Earth, the tides race ahead following Earth's 24-hour rotation. A tidal bulge tugs the Moon slightly ahead in its orbit, causing the Moon to accelerate. In this way angular momentum is transferred from Earth's rotation to the Moon, causing our length of day to slowly increase and the Moon to spiral outward.

LLRE reports the Moon's distance increasing by $3.82 \pm .07$ *cm/yr*. (The $\pm .07$ *cm/yr* is a standard deviation.) This rate is known to be "anomalously high". As calculated by Dr. Bruce Bills of NASA with Richard Ray, if the Moon were receding at this speed today, it would have occupied the same place as Earth barely 1.5 billion years ago. Studies of Apollo lunar samples, which I participated in, show conclusively that the Moon has existed separate from Earth over 4.5 billion years.

LLRE is a very precise experiment, checked exhaustively for errors, but it assumes that the Speed of Light is constant. Fortunately we have ways to measure the Moon's orbit independent of light. One method uses ancient sediments, called *tidal rhythmites*. Millions of years ago, tides washed up on an ancient shore. Each day the tides left a new layer of sediment, which became fossilized. By studying these tidal rhythmites, geologists can calculate the length of the lunar month and therefore the Moon's distance millions of years ago.

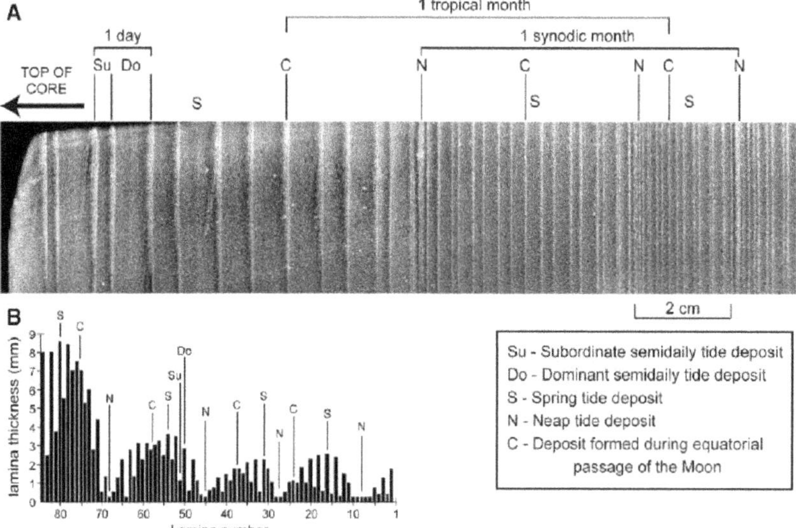

The Mansfield sediment of Indiana was formed 310 million years ago, when Indiana was beachfront property. Today the sediment survives in fossilized form, and has been studied by geologists. Mansfield's tidal layers indicate a lunar distance of *375,300 ± 1,900 km*. If we subtract this from today's *384,402 km* and divide by 310 million years, the Moon recedes at only *2.9 ± 0.6 cm/yr.* Other fossilized sediments also indicate the Moon receding more slowly than LLRE reports.

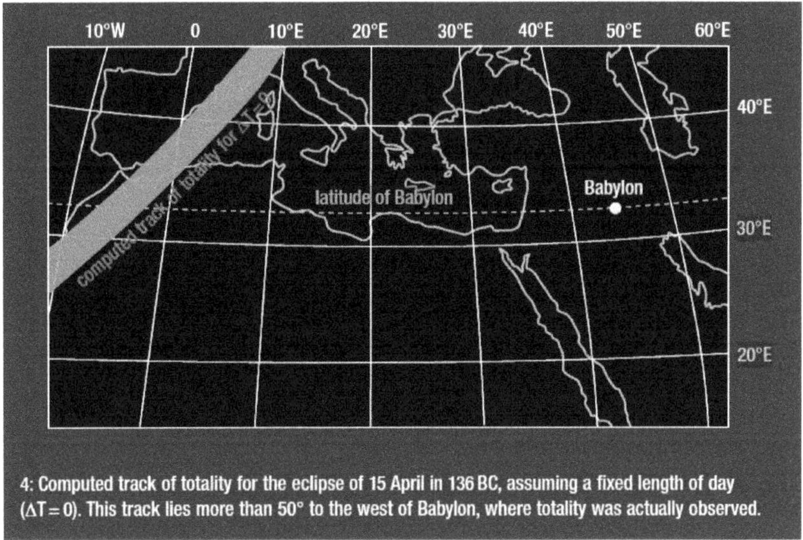

4: Computed track of totality for the eclipse of 15 April in 136 BC, assuming a fixed length of day (ΔT = 0). This track lies more than 50° to the west of Babylon, where totality was actually observed.

Another method of determining the Moon's recession rate comes courtesy of ancient astronomers. Babylonians recorded an eclipse as early as 721 B.C. We have other eclipse reports over thousands of years. The track of a total eclipse on Earth's surface is narrow and dependent on Earth's rotation rate. If an astronomer in Babylon reported a total eclipse on a day in 136 B.C., it provides a precise measurement of how Earth's length of day has changed. In turn this tells how much angular momentum has been transferred to the Moon, and how the lunar orbit has changed.

Dr. F. Richard Stephenson and Leslie Morrison in the United Kingdom spent many years cataloguing hundreds of historical eclipse observations. Eclipse records covering 2700 years show Earth's length of day changing by only *1.70 ± .05 msec/yr.* This indicates a lunar recession rate of *2.82 ± .08 cm/yr,* in agreement with sedimentary data. LLRE differs by more than 12 standard deviations.

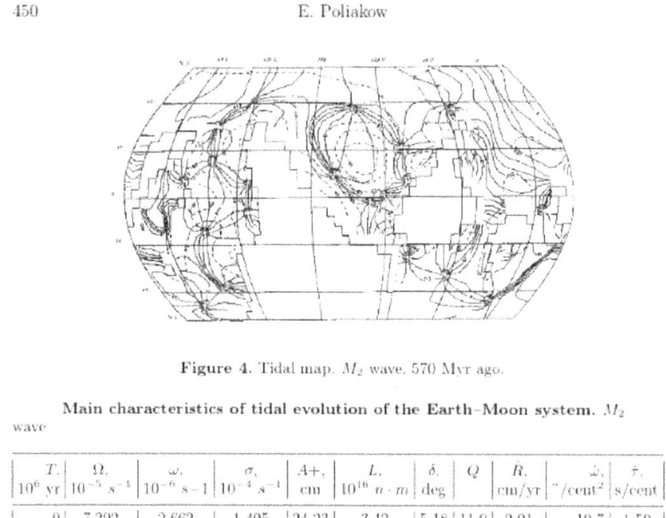

450 E. Poliakow

Figure 4. Tidal map. M_2 wave. 570 Myr ago.

Main characteristics of tidal evolution of the Earth–Moon system. M_2 wave

T, 10^6 yr	Ω, 10^{-5} s^{-1}	ω, 10^{-6} s−1	σ, 10^{-4} s^{-1}	A+, cm	L, 10^{16} $n \cdot m$	δ, deg	Q	$\dot{R}$, cm/yr	$\dot{\omega}$, "/cent^2	$\dot{\tau}$, s/cent
0	7.292	2.662	1.405	24.23	3.42	5.18	11.0	2.91	19.7	1.59

A third way to determine the Moon's recession rate is through computer simulation. The tidal interactions between Earth and Moon are subject to depth of the oceans, the location of ocean basins, and the movement of continents over time. At one time the continents were thought to be fixed; now we know their locations change. Dr. Eugene Poliakow of the Central Astronomical Observatory in Russia produced a very detailed computer model of the Moon's orbital changes.

The simulation was successfully used to predict ocean tides. According to Poliakow's simulation, today's recession rate is *2.91 cm/yr*, agreeing with sedimentary and eclipse data.

If these three independent experiments are given equal weight, their average value would be *2.88 cm/yr.* LLRE's laser light disagrees by *0.94 ± .07 cm/yr.* Mercury's orbit precesses by *5600 arc seconds* per century, but a discrepancy of only *43 arc seconds* per century was considered proof of General Relativity. When LLRE reports the Moon receding 1/3 faster than other experiments, over *12* standard deviations, it is a huge anomaly.

If the Speed of Light is slowing, time for light to return would increase each year, making the Moon appear to recede faster according to LLRE. $GM=tc^3$ predicts the size of the lunar anomaly to be *0.935 cm/yr.* If the Moon's distance were fixed, LLRE would still report it receding at *0.935 cm/yr* due to a slowing Speed of Light. This prediction fits the anomaly to less than 1/10 of a standard deviation, a scientific bulls-eye!

This is one of the most accurate scientific predictions ever made. Errors in results are measured in standard deviations, called Sigma. Physicists say a *5-Sigma* result is proof; this is a *12-Sigma* result! The chances of it being coincidence are vanishingly small, one in many millions.

Like Jupiter's moons and Mercury's orbit, exploration of Earth's Moon gives clues about the Universe and light. As *c* change would make the Universe appear to accelerate, it would also make the Moon appear to recede faster.

Together these observations show why this is an exciting time to be in science. Signs of change in the Speed of Light come from the microwave background, distant galaxies, supernovae, the Sun and Moon. Mysteries that seemed unrelated may be connected. In some cases there are alternate explanations available. It is possible to believe that the early Universe inflated faster than light, if one ignores the data.

Supernova redshifts have caused speculations of a repulsive "dark" energy. Someone can imagine that an unknown force has finely tuned Earth's temperature for our comfort. These explanations share the quality of being invisible and unobserved in the lab. They also share the supposition that we are the centre of the Universe, between extreme heat and cold or between repulsive forces driving expansion. Starting from a first principle is like knowing all the answers in the back of a book.

At Johnson Space Center we had responsibility to support the International Space Station. ISS orbits a spherical Earth, following Newton's laws of gravitation. The Station's solar panels convert sunlight into electricity, an application of Einstein's photoelectric effect. By exploring over the horizon, we may yet find more clues about the shape of the Universe.

Sunrise and the International Space Station

ACES: A Better Clock

In addition to a more distant hilltop, we can also complete Galileo's experiment with a better clock. The Atomic Clock Ensemble in Space is scheduled be installed aboard ISS in early 2022. In microgravity an atomic clock is more precise than on Earth, because Earth's gravitational pull does not interfere with the vibration of atoms.

An international project led by the European Space Agency, ACES will be orbited aboard a SpaceX Falcon 9 and placed by the Station's robotic arm outside the European Columbus module. One of ACES' chief science goals is to search for "anisotropies" or changes in the Speed of Light. Initial mission length is 18 months, with possible extensions to 3 years or more. At this writing, ISS is funded until at least the year 2024.

ACES could potentially measure a change in c smaller than *1 in* 10^{10}, or *3.00 cm/sec*. Since light would slow this much in 4 years, hopefully atomic clocks will enjoy a long stay aboard ISS. Will the Space Station's atomic clock verify a "*c* change" in physics? Time will tell!

Evidence that *c* has changed in time comes from all around, from galaxies, supernovae, the Sun, Moon and possibly the Space Station. Eventually this tide of evidence may be too big to ignore. Many mysteries of the Universe are explained if we open our eyes to light.

Next: Tour of the Universe

9

Tour of the Universe

"A single candle can both defy and define the darkness"
—Anne Frank

Italy seen from ISS

If any pleasure approaches that of scientific discovery, it can be found piloting an aircraft. Flying high above the Earth puts many problems in perspective. Cities at night are ablaze with a million lights. From experience we know that the pools of light are just a hint of the mass below. Between those lights are roads, buildings, and people going about their business. The majority of mass lies hidden in the darkness.

When they have nothing better to do, scientists speculate about multiple universes. There is another Universe, occupying the same Space/Time but hidden from our eyes. Theory predicts, and observations confirm, that the mass we can see is just *4.507034%* of the total. To be aware of the other *95.49%* is to be a seeing woman among the blind. Here we will take a quick flight around the Universe, searching for what lies in the darkness. We cannot know all the locations of invisible mass, but here are some good places to look.

The White Mountain

Even a journey into Space should begin at home. The Big Island of Hawaii rose from the sea barely one million years ago, product of an ancient volcanic "Hot Spot." This plume of hot lava begins deep within the Earth, near the boundary between core and mantle. The islands sit atop the Pacific Plate, which has been slowly inching across the top of the plume for over 70 million years. The northwest islands, including Kauai and Midway, are the oldest. The Big Island is the youngest and most volcanically active. The volcanoes are an exciting place to study Earth's formation, and a connection with Black Holes.

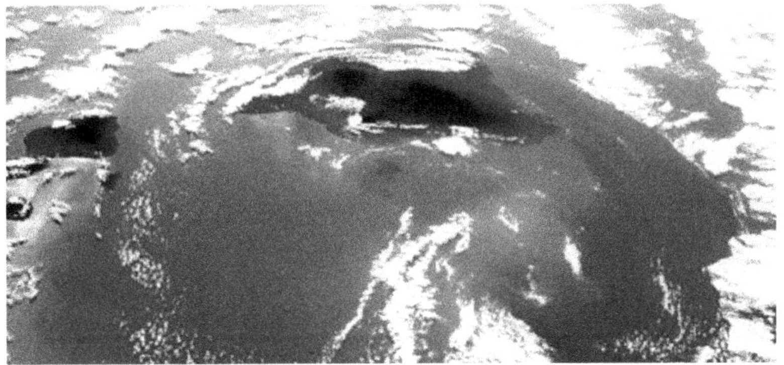

This is the Big Island as photographed from a Space Shuttle. To the left is Maui and at top is the Northeast coast including Hilo. The Pacific trade winds blow from East to West. Clouds pile up along the coast like waves against a ship's prow, then part to leave an island wake. Hilo is the wettest city in the US, with over 100 inches per year. It is a gardenlike setting surrounded by streams and waterfalls. One reason that Mauna Kea is a good observing site is because the air arrives "clean" without turbulence from another land mass.

Our Big Island's highest point is the dormant volcano Mauna Kea, "White Mountain" in Hawaiian. It's summit, 4,205 metres above sea level, pokes above the clouds and much of the atmosphere's water vapour. From ocean floor to summit Mauna Kea is 10,200 metres high, taller than Mount Everest. Today the summit is studded with astronomer's telescopes, for Mauna Kea is one of the best places on Earth to look at the sky.

160°W
159°W
158°W
Hanalei
Ni'ihau
Pu'uwai
Kekaha
Kapa'a
Lihue
Kaua'i
Kauai Channel
Ka'ula
Kahuku Point
Waialua Bay
Laie
0 5 10mi
0 5 10 15km
Kaena Point
Waialua
Haleiwa
O'ahu
Wahiawa
Kaneohe
Mokapu
Pearl
Kailua
Barbers Point
Honolulu
Kure
175°W
170°W
165°W
160°W
Midway
Pearl and Hermes Reef
Lisianski
Maro Reef
Laysan
25°N
Gardiner
Northwestern (Leeward) Islands
French Frigate Shoals
Necker
Tropic of Cancer
Nihoa
20°N
0 200 400mi
0 250 500km

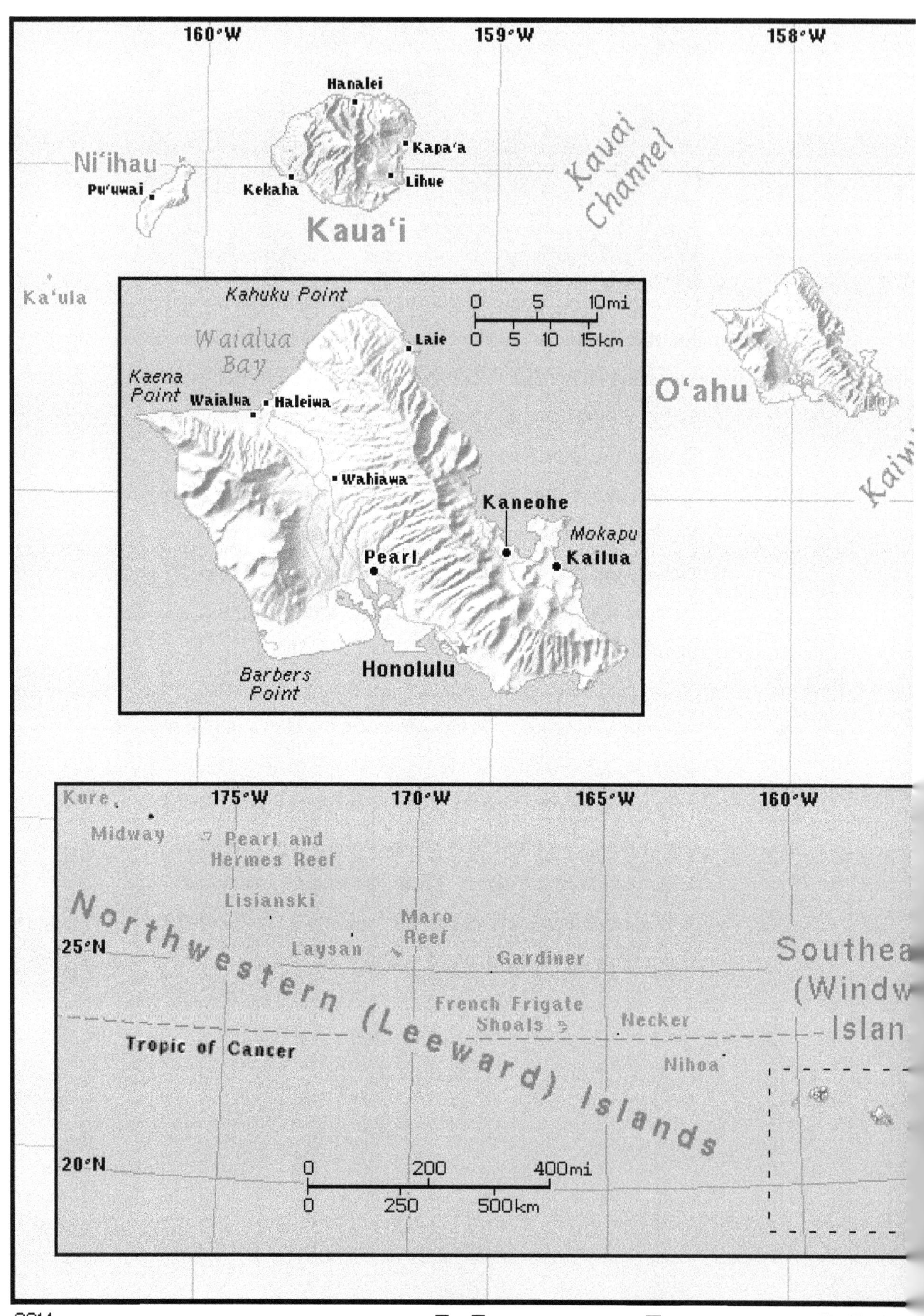

2014

H A W

157°W
156°W
155°W
22°N
21°N
20°N
19°N
Channel
Moloka'i
Kalaupapa
Maunaloa
Kaunakakai
Kahului
Lahaina
Wailuku
Lanai City
Maui
Hana
Lana'i
Kaho'olawe
Alenuihaha Channel
Upolu Point
Hawi
Hamakua Coast
Honoka'a
Kawaihae
Waimea
Kea
4207m
Hilo
Keahole Point
Kailua
Mtn View
Loa
4169m
Pahoa
Cape Kumukahi
Hawai'i
Capt Cook
Kilauea
Kona Coast
The Big Island
Kaimu Beach
Pahala
Miloli'i
Punalu'u
Na'alehu
Ka Lae
0
50
100mi
0
50
100km
155°W
stern
ard)
ds

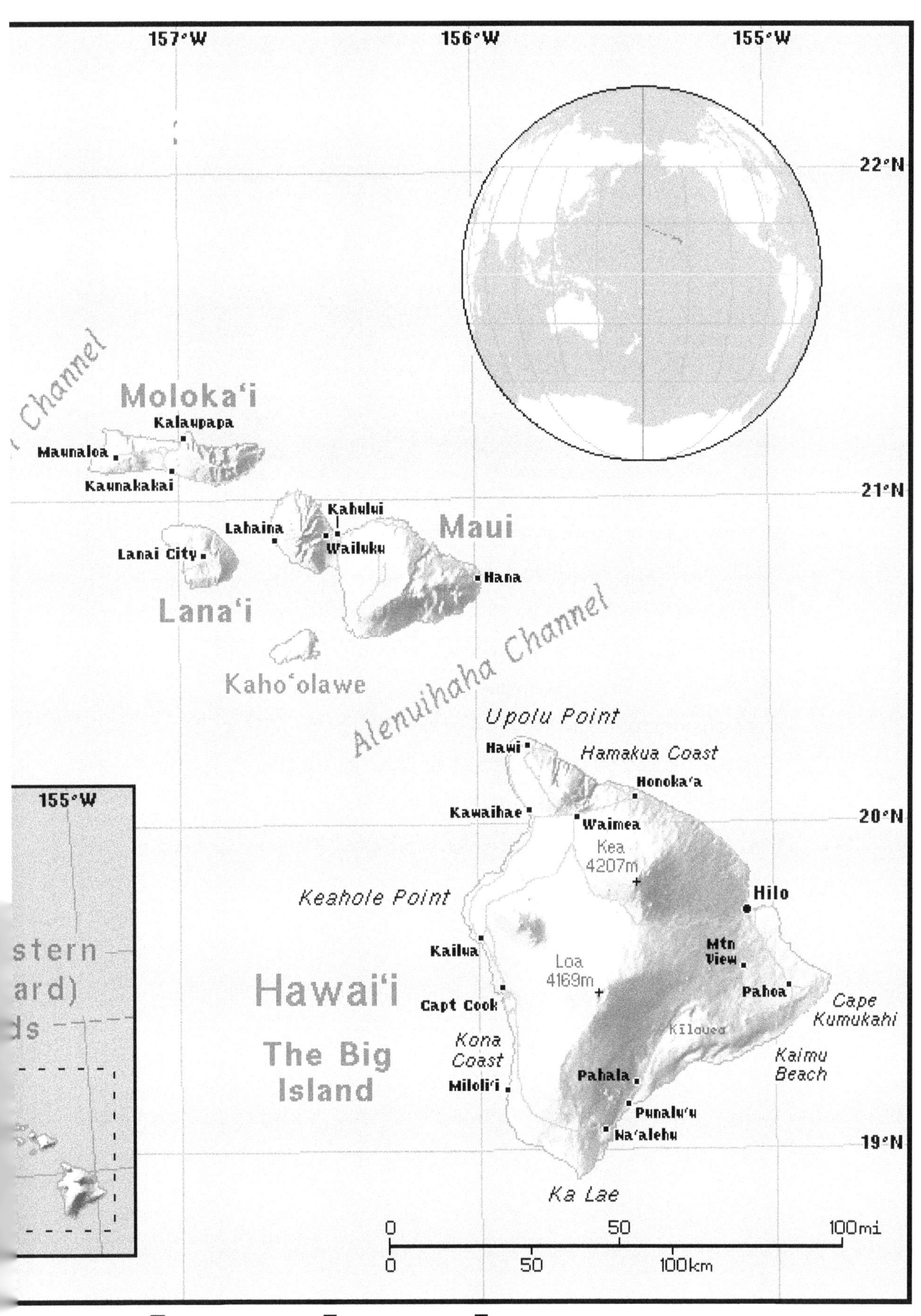

A I I

The mountaintops of the Big Island have long been considered sacred, with Mauna Kea the most sacred of all. Traditionally only high-ranking chiefs and priests were allowed on the summit. The remoteness and altitude make it a difficult place to reach. I waited a number of years to visit the summit because children under 16 are not allowed. Astronomers may spend the night on the summit, but only in groups of at least two. The rarefied air is not for the faint-hearted or anyone with respiratory issues. Most astronomers stay at *Hale Pohaku*, the lodge at the 2700-metre level.

Mauna Kea is home to 12 major telescopes.

It was approved as site of the Thirty Meter Telescope. The most famous are the twin 10-meter telescopes of the Keck Observatory. The next biggest apertures belong to the 8.3 metre Subaru telescope, the world's largest single-mirror telescope; and the 8.1 metre Gemini North telescope, which can see in both visible and infrared light. The Canada-France Hawaii Telescope (CFHT) also looks into the infrared sky. Part of the summit is named "Millimeter Valley" for observatories that peer into the ultraviolet and radio frequencies. These millimeter waves allow astronomers to peer through some of the dust that fills our galaxy.

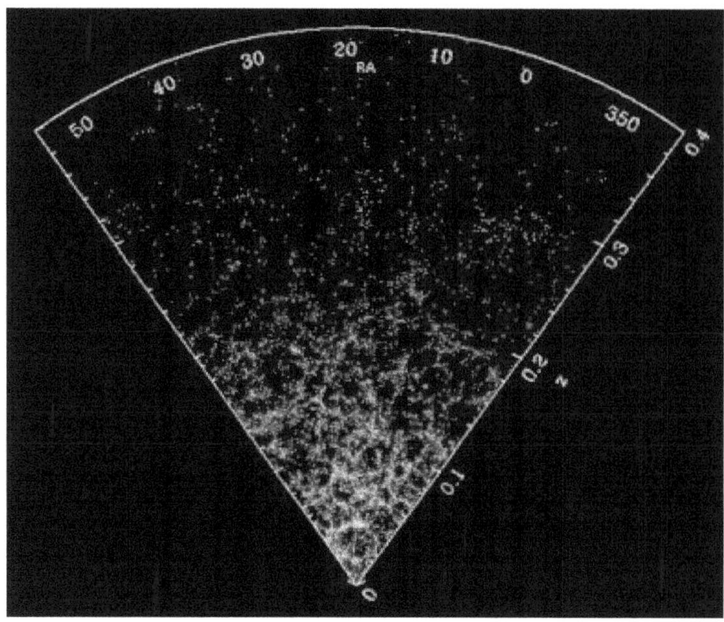

Astronomer Margaret Geller and my friend the late John Huchra mapped galaxies since 1985. Today the Sloan Digital Sky Survey has mapped a slice of the visible Universe. Galaxies and clusters form enormous sheets stretching across millions of light-years. These great walls are boundaries of vast bubbles containing most of the Universe's volume.

A fish in the Barrier Reef avoids dark holes in the coral, because they could hide something that could eat her! One hopes that humans are more intelligent than fish. It would be foolish to think that humans know everything in the Universe, or to assume that the "voids" are empty. Something may indeed be in those dark places, something hungrier and more massive than humans have imagined.

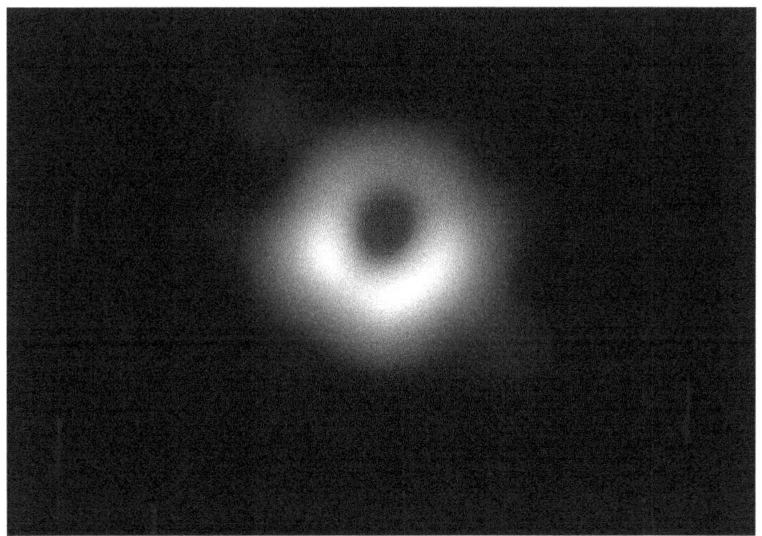

Powehi the Black Hole

The Speed of Light predicts that quantum fluctuations near the Big Bang could have created singularities of almost unlimited size. The largest would be true Black Holes devouring everything within reach. In the early Universe they would have cleared whole regions of matter. Models predict that about *68.3%* of mass in the Universe would end up in such regions. Since the Cosmic Microwave Background dates from 380,000 years after the Big Bang, this majority of mass would not show up in the CMB. The fragile spaces in between, balanced in a tug-of-war between masses, would form sheets where galaxies could form. Magnified by expansion of the Universe, the picture would look exactly like what Sloan has seen.

Despite their immense size, ultra-massive Black Holes would be very difficult to detect. Their intense gravity would devour any light or radiation. They would, however, produce magnetic fields that could be detected. Astronomers have found such powerful magnetic fields, the source of which has been a mystery. Black Holes can also be located through gravity. One such object is already drawing us inexorably toward it.

The Milky Way is within a Local Group of about 30 galaxies. This group is part of a Local Supercluster containing dozens of clusters spread over 100 million light-years. At the centre of this is the Virgo Group, which contains thousands of galaxies. Everything in this vast community is being drawn at *600 km/second* toward something in the direction of cluster Abell 3627. This immense mass, about 10^{16} times the mass of our Sun, has never been seen. Its discoverers have dubbed it The Great Attractor. This object is not alone; there is evidence of another Attractor hidden beyond it. We only know of the Great Attractor's existence because it influences our galaxy directly; the Universe could contain many others.

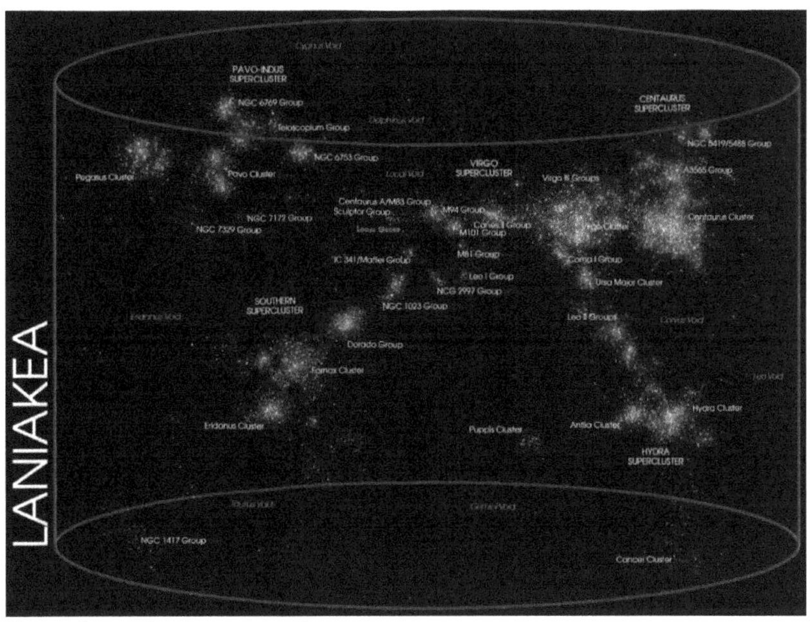

Laniakea, the immense heaven

During 2014 astronomers at University of Hawaii, using a new technique for determining distances, reported that our Virgo Cluster is just a small part of an enormous supercluster This immense structure has been named Laniakea, Hawaiian for “immense heaven”. Laniakea contains 100,000 times the mass of our Milky Way galaxy and spans 500 million light years. The galaxies within Laniakea, including ours, are slowly being drawn toward the Great Attractor.

There are other ways to detect what cannot be seen. This photo is from the Hubble Telescope Deep Field. General Relativity predicts that gravity bends light rays. Abell 2218 is a cluster of galaxies, each containing billions of stars surrounding a massive Black Hole. Around the edges we see galaxies shifted into red. These galaxies are not located around the cluster but behind it, 5-10 times more distant. Their light has been bent by gravity, traveling around the cluster to reach us! The images of galaxies have been stretched into arcs, as if seen through an immense lens. This phenomenon is called gravitational lensing. This bending of light is a confirmation of General Relativity.

With the aid of computers, we can use gravitational lensing to see where the mass really lies. The cluster's density peaks at the galaxies, but is even higher toward the centre. Most of the cluster's mass is dark. Something in the centre holds the cluster together and bends the light from more distant objects. Whatever lies there must be far more massive than the galaxies, and concentrated in a small region. We could be seeing the lair of an ultramassive Black Hole. This object did not form a galaxy because it swallows all light and matter in its vicinity.

The European Space Agency's XMM-Newton spacecraft took this X-Ray photo of Cluster RXJ0847.2+3449. We see this cluster of galaxies as it was about 7 billion years ago, when the Universe was half its present age. A team led by Alain Blanchard of the MidiPyrenees Observatory in France has found many such clusters. They help confirm whether "dark energy" exists.

If repulsive "dark" energy filled the Universe, 5 billion years ago it would have begun to dominate. Formation of galaxy clusters would have slowed as repulsion overcame gravity. We would see more clusters in the past than exist today. Astronomers have found significantly fewer clusters in the past, telling us that clusters are still forming. Rather than tearing itself apart with "dark energy," the Universe continues to form more complex structures.

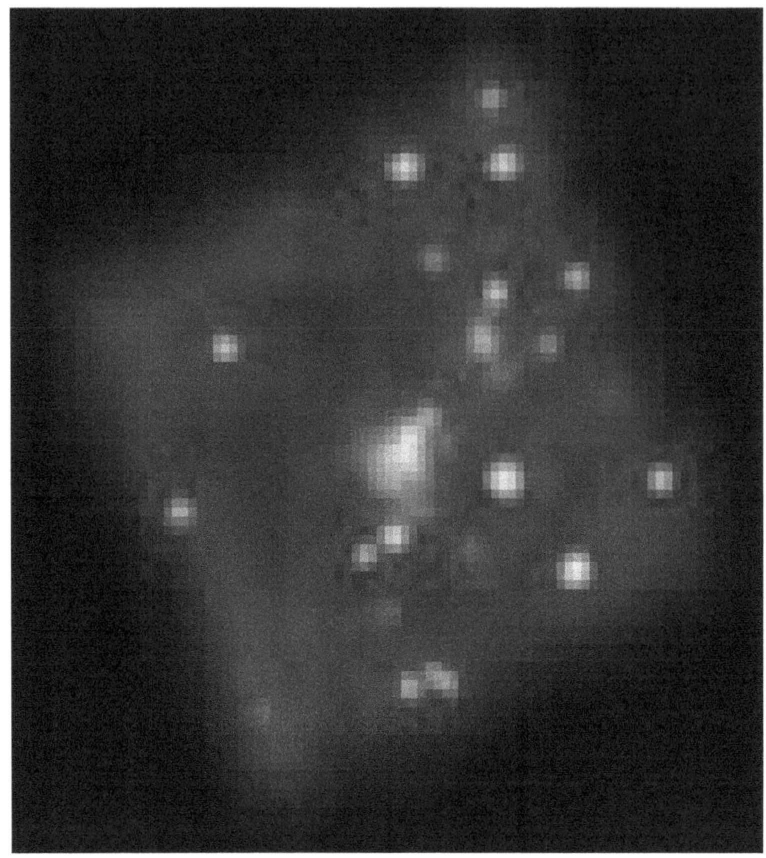

Results from XMM-Newton, verified by the Chandra X-Ray Observatory, show that gravitating "dark" mass is four times as abundant as previously thought. Instead of just 24%, it comprises the 95.49% of the Universe that is not baryons. It is by far the majority of mass in the Universe, where our matter is just a footnote. There is no need for a repulsive "dark energy" to fill the rest. We are the lights hinting at something else in the darkness.

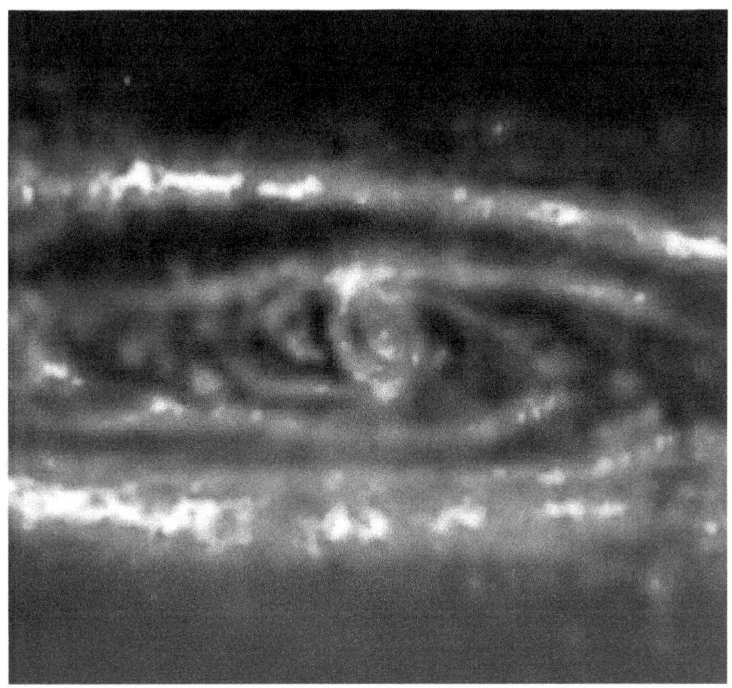

The Hubble expansion does not affect all galaxies equally. Andromeda, the nearest galaxy, is moving toward us. The Spitzer Space Telescope made this infrared image. 200 million years ago galaxy M32 collided with Andromeda, leaving huge ripples in Andromeda's disk. In another 5-10 billion years Andromeda and our Milky Way will collide.

The two galaxies do not contain enough mass to cause this attraction. There is invisible mass between them, about ten times the mass of a galaxy. Andromeda's core is a Black Hole with about a million times our Sun's mass. Not big enough to swallow everything, it may have been the seed around which the galaxy formed.

Astronomer Fritz Zwicky thought that "dark matter" surrounds and fills the galaxies. By measuring the velocities of stars at various distances from galactic centres, Vera Rubin showed that galaxies contain far more than meets the eye. After three men who claimed "dark" energy received a Nobel Prize in 2011, Vera Rubin passed away in 2016 at age 93, without being awarded that prize.

Cosmologists, like Stephen Hawking, believe that the Big Bang created billions of tiny Black Holes. Attracted by gravity to larger objects, they would form haloes around the galaxies. Exactly like "dark matter," they would exert gravitational influence but emit no light. Models predict this mass at *23.87%* of the total, a figure confirmed by WMAP. The term "dark matter" may be misleading--these Black Holes may have never been matter.

Globular cluster Messier 80 contains hundreds of thousands of stars and orbits around our Milky Way. Harold Shapley used observations of these objects to locate the Milky Way's centre. A globular cluster contains some of the oldest stars in our galaxy. Astronomers do not know how the globular clusters formed so early, or what holds them together. One way to form them would be from a medium-sized Black Hole orbiting the Milky Way. Since this Black Hole formed primordially, it would have been a magnet for early stars.

If our galaxy is orbited by Black Holes, have they collided with the galaxy? The result could be us!

One of the most spectacular pictures from the Hubble Space Telescope is of the Eagle Nebula. The immense pillars are cocoons containing infant stars. Somehow the highly diffuse interstellar gas is collapsing into a point so hot and dense that nuclear fusion is ignited. During this process, something prevents that heat from dissipating the cloud and ending the birth.

One way to produce this formation would be if a tiny Black Hole, or many, collided with the cloud. They would not be big enough to suck up the whole cloud, but would leave behind pillars of gas like pellets fired through cotton candy. At their centres, gravity and radiation would produce the conditions for stars to ignite. The heat produced would balance gravity's inward pull so that the stars burned steadily. The Black Holes may still be in the stars' cores, quietly contributing to their power output.

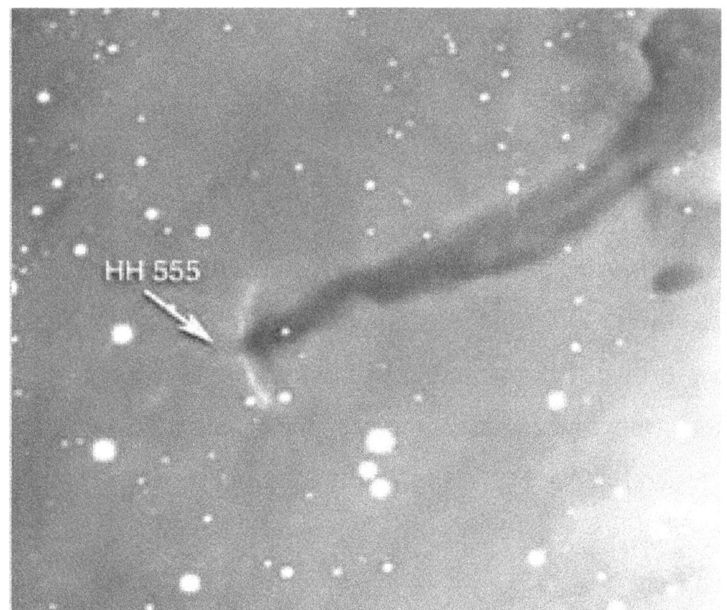

Thanks to the Mayall 4-meter telescope atop Kitt Peak in Arizona this is HH 555, a Herbig-Haro object in the Pelican Nebula. At the pillar's tip gas is drawn into a disk, which will form a baby star. At the top and bottom are two bright jets, guided by magnetic fields from the core. Twin jets and a magnetic field are telltale signatures of a hidden Black Hole.

Theories of the Sun have advanced over time. As late as the 1920's most astronomers would lecture that our Sun was made of iron, and glowed in the sky like a hot poker. Only a young astronomer named Cecilia Payne suggested that the Sun's spectral lines could be interpreted as hydrogen. Possibly because Payne was a woman, her ideas were dismissed.

Though Cecilia had completed her studies at Cambridge, the University did not award degrees to women. The equations of nuclear fusion were still being worked out, and most scientists doubted that Black Holes existed. Eventually the young woman was vindicated. As our knowledge advances, so must theories of the Sun.

This is a sunset photographed from the east coast of the Big Island. A Black Hole could conceivably exist in the second last place we would think to find one, inside the Sun! No physical law prevents this; most laws of astrophysics would remain unchanged. A Black Hole would feel right at home in the radiation and pressure of a stellar interior.

If a Black Hole had memories, it would remember the conditions of its birth near the Big Bang. Presence of a tiny singularity, which is predicted by cosmology, would ignite the nuclear reactions within a star. The primary power source of stars would then be fusion, occurring in the immense heat of stellar cores. The Black Hole's rotation would cause the star's inner layers to rotate faster, contributing to a magnetic field. Astronomers now know that our Sun's core rotates faster than the outer layers due to some mysterious influence. A Black Hole could be literally in front of our face each morning.

Niels Bohr is best known for describing the heart of an atom, but he also turned his attention toward stars. Bohr's atomic model is based upon a miniature Solar System. In the early part of the 20th century many old ideas of science were questioned. Bohr would say that "Your theory is crazy, but it's not crazy enough to be true." In work that was never published, but related to his friend George Gamow, Bohr hypothesised that a star's heart is made of an unknown material radiating at a constant temperature. The region surrounding this core would be composed of hot plasma. If temperature of the core were greater than that of the plasma boundary, an equilibrium would be reached. The star would maintain a constant luminosity.

Bohr's hypothesis required a material radiating at a constant temperature, something unknown in the 1930's. Though Black Holes are predicted by Relativity, even Einstein himself didn't believe they exist. In 1974 Stephen Hawking reached the amazing conclusion that they radiate at a temperature depending upon their mass. For many Black Holes this temperature would be nearly constant. Connecting Bohr with Hawking, a Black Hole at a star's heart would be just the thing to maintain a constant luminosity.

Only occasionally would Black Holes within stars reveal themselves. In their twilight years, the largest stars would consume all their fuel until fusion abruptly stopped. The equilibrium between outward pressure and gravity would end, causing a catastrophic collapse. A star's mass suddenly falling into a Black Hole would produce an immense explosion, a supernova.

A supernova leaves in its aftermath a neutron star, composed of matter so dense that a spoonful would weigh a ton. Some neutron stars are pulsars, emitting spinning beacons of radiation like a lighthouse. Cause of these jets has been another mystery. The axes of the jets do not coincide with the neutron stars' axis of rotation, indicating that something else inside the neutron star creates the jets. The pulsar's jets are exactly like the twin jets from a Black Hole. Other neutron stars are *magnetars*, objects with magnetic fields so powerful that they defy Maxwell's equations to describe them. Once again, jets of radiation and a magnetic field are telltale signatures of a hidden Black Hole. It would rotate independently inside a neutron star, producing both a spinning beacon and magnetic field.

A journey ends with a return home. We began at our own planet, by seeing how people found that Earth is round. From there we have visited the solar system, the stars, the galaxies and even the beginning of the Universe. Now we can see our home with new eyes. One of the biggest surprises of cosmology may wait beneath our feet. It was created shortly after the Big Bang and has been waiting since before life began for our discovery. Our planet's interior contains enough mysteries for another book, about a Hole in the Earth.

Next: A Theory for Everything?

10
A Theory For Everything?

"In the midst of darkness, light persists"-Mahatma Gandhi

Starting from a child's first explorations, we have wandered from the beginning of the Universe and back to our home planet. Science has discovered that similar rules apply to the whole Universe. In many ways, every part of the Universe resembles every other. Today we may be able to determine the size and shape of the Universe. Our seemingly complicated Universe may be described in a simple equation.

Our position today is like that of humans thousands of years ago, standing on the shore wondering about Earth's shape. Some said that it was flat, others looked at evidence and concluded Earth was spherical. Eventually the curved theory won out. Today we are barely able to find evidence for curvature on the Universe. With more data, our children will understand its size, shape, and the simple principles it is based upon.

A spherical Universe, resembling Earth or a mother's womb, appeals to our sense of aesthetics. The simplicity of spherical shapes appealed to Pythagoras, Pascal, Einstein and even Edgar Allen Poe. If someone thinks that the Universe is spherical, they are in very good company. Today their intuition has been borne out by mathematics.

Topology is the study of surfaces. In topology, shapes can be stretched like dough and still be mathematically similar. A sphere is similar to a cube in that one shape can be stretched into the other without tearing any holes. A donut and a teacup are similar in that they each have one hole.

From tiny cells to stars, 3-demensional objects tend to form spheres. Henri Poincare was the father of modern topology, the study of surfaces. In 1904 he conjectured that the most natural form for a 3-dimensional space is the surface of a sphere in 4 dimensions. Poincare could not prove his conjecture, and it remained unproven for a century.

Our astronauts experience a simple form of Poincare's Conjecture every time they venture outside on EVA. 2-dimensional sheets of fabric and rubber have been made into human-shaped spacesuits. Upon emerging into the larger world of Space, pressure tends to balloon the spacesuit into the shape of a sphere. The Spacesuit designers and the astronauts must fight this tendency to form spheres, making movement during EVA very difficult.

Grigorii Perelman was born in St. Petersburg, then called Leningrad in the former Soviet Union. His father, an electrical engineer, encouraged an interest in science and gave his son many books. His mother, a math teacher, took

Grigorii to operas and encouraged him to enjoy music. Perelman showed an early talent for maths, attending a specialized school for mathematics and physics before entering Leningrad University at age 16. After graduation, he took a post at the Steklov Institute of the USSR Academy of Sciences. After the collapse of the Soviet Union, Perelman came to the US and found work at American universities. In the United States he was exposed to new worlds of mathematics, including Poincare's mystery.

Mathematician Richard Hamilton earlier had proposed a solution based upon Ricci flow. This flow insures that the Ricci curvature tensor, important in General Relativity, is always positive. Heat introduced into a tub of cold water will diffuse out until the water has a uniform temperature. Hamilton suggested that curvature could also diffuse out until a surface of equal curvature, a sphere, was achieved. Proving this in 3 dimensions seemed an insurmountable problem. Perelman met Hamilton while in the US, then returned to Russia in 1995. For the next 7 years he would focus almost exclusively on proving the Poincare Conjecture.

By the turn of the millennium, technology was at the point where scientists could work in isolation and still communicate ideas. Scientists from Aristarchus to Galileo wrote their own books to spread ideas. Einstein first published in journals, which had to pass review by suspicious editors. In November 2002 Perelman began posting outlines of a proposed proof on the internet. Though some mathematicians realised he was on to something, parts of the proof were incomplete. By Spring 2006 the proof was complete, and the world realised what had been accomplished.

Perelman's achievement made him eligible for multiple maths prizes, but he has not bothered to collect them. Like many great minds, he prefers to work in privacy. Perelman considers himself retired from the math field, and takes his greatest pleasure in the opera. He prefers the inexpensive seats high in the opera house. Away from distractions, he can listen for the harmonies in singer's voices. Like Pythagoras, Perelman enjoys both music and maths.

The Poincare Conjecture has applications to our Universe. It implies that the only possible shape for a 3dimensional Universe is the surface of a 4-dimensional sphere. Einstein calculated that if density of the Universe were great enough, gravity would curve it into such a sphere. This shape can be maintained without collapsing if there is a certain density. That density is not "critical" but the stable density. Mathematically this density is an "attractor," pointing to just one possible Universe. If the Universe were created from nothing, then mass would be created to push the Universe toward this density. Since it is possible to describe the Universe with one equation, others have breathlessly speculated about a "theory of everything."

Describing even the form if such a theory would be very difficult. Einstein, in his later years of fame, worked fruitlessly to find a "unified field theory" combining gravity and electromagnetism. Many good minds have been wasted trying to unify General Relativity with Quantum Mechanics. It may be a waste of time trying to create an all-encompassing theory of everything. Physicists would oppose such a theory, for it would put many of them out of work.

Small links between General Relativity and Quantum Mechanics already exist. Stephen Hawking's prediction of Black Hole radiation uses both. We have seen that the Speed of Light c, which is a part of Relativity, and the Planck value h of Quantum Mechanics may be linked. Instead of an overarching theory of both, QM and GR may be stitched together from small links.

As promised, there is only one equation in this book. However, an equation may have more than one form. Our units, like meters and seconds, are based on arbitrary values. An alien species would use entirely different units. Max Planck in 1899 suggested a system of units based on combinations of his value h, the Speed of Light c, and Newton's gravitational constant G. The "Planck units" of mass, length and time are made up of combinations of these values. These tiny units may have significance for our big Universe.

The Only Equation In the Book, when combined with an expression for the Universe's scale *R*, in Planck units becomes even more simple:

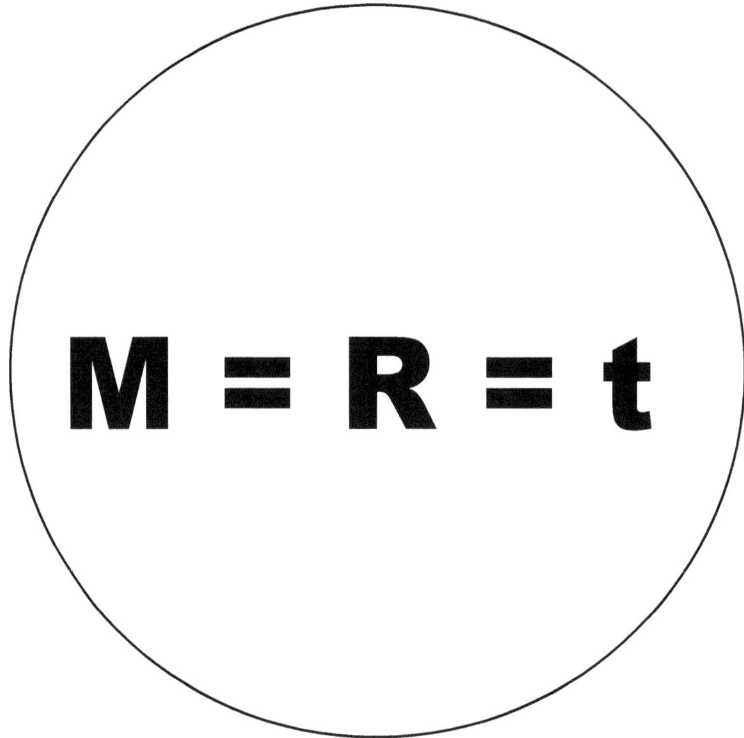

Mass *M* of the Universe, about 10^{61} Planck masses, equals its scale *R* (10^{61} Planck radii) equals its age *t*. We may call this M-Theory, or an Equation of the Universe.

The number 1 followed by 61 zeroes, like the Universe, is very large. Arthur Eddington, whose eclipse expedition tested Einstein's General Relativity, thought that large numbers like these were significant. In his later years he used them as the basis for a "fundamental theory". This was one theory that has never been proven.

Paul Dirac, who predicted the existence of antimatter, thought that electromagnetic force, gravitation, size of the Universe and size of an atom were all related. Dirac made the prediction that the gravitational constant G changed over time. This theory has also never been proven. However, Dirac's problem with large numbers is also solved when $GM=tc^3$.

Nature contains many values and parameters. These include the masses of the proton and electron, their electric charge, and other measured quantities. Numbers without dimension include the fine-structure value α and various ratios found in nature. The standard model of particle physics contains many "free parameters" that can be measured but not yet predicted. This book has shown how one important value, the Speed of Light, can be calculated at different times. In the future other values and supposed constants may be calculated from pure mathematics.

The desire to understand the Universe is part of humanity. It is a natural extension from crawling, walking and climbing. To fly is to reach and explore far more places. Every human has at one time or another dreamed about flying. Consciously we have imagined angels and superheroes that resemble us but can fly. Though humans have no wings, we are given minds that can dream and organise. The ability to dream has enabled us to build flying machines and spacecraft. Humans are the only species on Earth that can leave the planet.

The Universe continues to expand and grow more complex, as it has for billions of years. It will do so for countless hundreds of billions of years more. Just as a time near the Big Bang seems like a beginning to us, our time will someday seem like a beginning. Like children first opening our eyes to light, we are only beginning to understand the Universe.

11

Afterword: A Time of Light

"May it be a light to you in dark places,
when all other lights go out."—J.R.R. Tolkien

The afterword in this book's previous edition told of the headline-making events of 2014, when "smoking gun" evidence of an old inflated idea was found to be dust seen in the telescope. The idea of a flat Universe made its last inflated gasp before bursting. The year 2015 was designated by the United Nations as International Year of Light, 100th birthday of Einstein's General Relativity and 50th anniversary of the Cosmic Microwave Background. The past 5 years have been even more exciting.

In 2019 measurements of the Hubble value, which gives the age and expansion rate of the Universe, disagreed. Supernova researchers, the same ones who claimed "dark" energy, gave one value, while measurements of the Microwave Background from the early Universe disagreed. The discrepancy was so glaring that it was a "crisis in cosmology". If the Speed of Light had changed since the early Universe, the "crisis" would be easily explained.

Many believe there is trouble with physics. There have been precious few discoveries in the past 45 years. Physics has been mired by inflated ideas of a flat Universe and ideas that are not even wrong. The Speed of Light solves the problems of both "inflation" and "dark" energy.

The Only Equation In the Book, when written in the units of Max Planck, is so simple that it must be repeated.

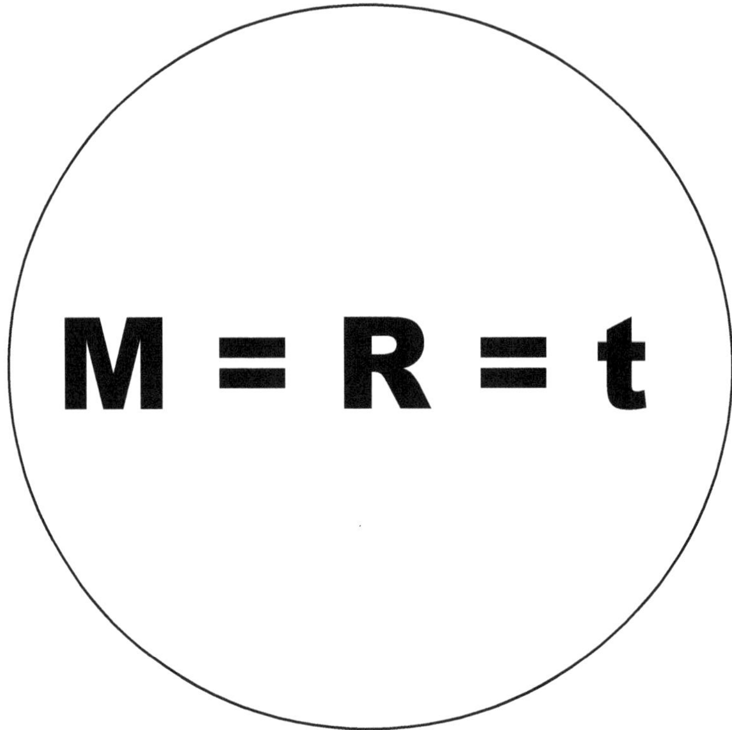

This expression is so evocative that life on Earth evolved for billions of years before figuring it out.

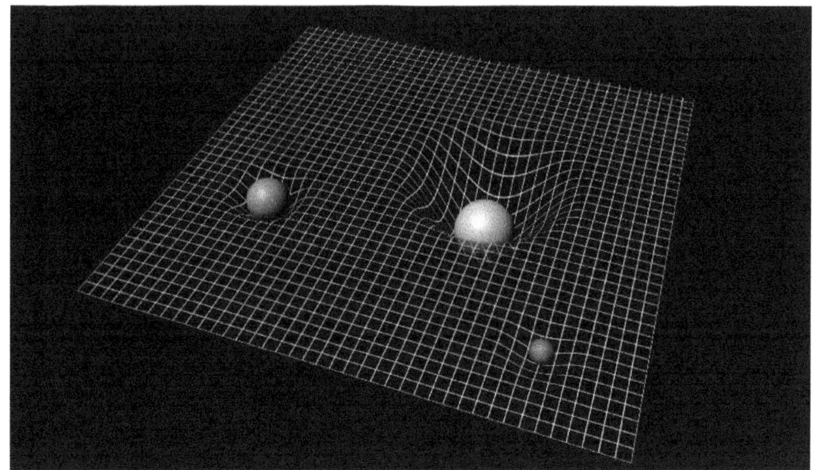

What does it mean? It hints that the Planck units of mass, length and time could be fundamental. Our giant Universe can be expressed as a multiple of these tiny units. The Universe may seem big and dark, but its size is something the mind can comprehend.

If the Planck units are fundamental, then the Space/Time we live in may not be continuous. Time would flow in tiny steps like the ticking of a clock. Space would be divided into tiny quantum bits, like a grid or a honeycomb. The Planck time, about 10^{-43}seconds, and Planck length of about 10^{-35} meters, are too tiny to measure.

In contrast the Planck mass, about $2.2 \times 10^{-8} kg$, is something we can see with our eyes. It is about the mass of a flea's egg or one of your eyelash hairs. While the size and age of the Universe increase over time, the Planck mass alone is predicted to be fixed. This mass may be a real constant in nature, a link between gravity and quantum mechanics.

Galileo using the Leaning Tower of Pisa showed that two objects of different mass fall toward Earth at the same speed, the experiment that Dave Scott performed on the Moon. Isaac Newton showed that this is duc to laws of gravity. Albert Einstein showed that objects fall because Earth's large mass causes Space/Time to be curved. Today these laws of gravity may need to be modified.

Many women and men have been employed dusting off the shelves. Dust drifts toward the floor because of Earth's gravitational mass, but dust particles do not attract each other gravitationally. Even when dust particles are left on the floor for a long time, gravity does not make them clump together. (The "dust bunnies" that gather in corners are a result of electrostatic and aerodynamic forces.) This is a clue that gravitational mass may be *quantized* at the Planck scale.

Two tiny particles (of less than Planck mass) are attracted to Earth because of Earth's large mass, but do not attract each other because gravitational mass is quantized. This quantum nature of gravitational mass is also simple, but something that has never before been tested. Many people have seen dust fall, as many saw apples fall before Isaac Newton saw a law of gravity. This is a concept that people who have spent their entire life in Earth's gravity have difficulty visualising.

We can demonstrate quantized gravity in a simple experiment that any laboratory or classroom can repeat. We start with two flea eggs, or small particles of less than the Planck mass. Spherical grains of limestone, called *oolites*, are a good test subject. We choose two with diameter of less than *0.2 mm*, about the size of the period at the end of a sentence.

The two particles are placed about *2.0 mm* apart on a low-friction surface such as Teflon. They are grounded to the surface to keep them from developing an opposing electric charge. If possible, the experiment is placed into a vacuum chamber to prevent air pressure from interfering with the experiment. Johnson Space Center contains a huge vacuum chamber, but any small vacuum bottle will do.

According to old laws of gravity, the two particles should attract each other. After many hours of observation, no movement is detected between them indicating that their gravitational mass is *quantized.* The experiment would work even better in the microgravity of Space, where the two particles would not rest on anything that would cause friction. If we remove frictional, electric, or aerodynamic (air pressure) forces, two small particles will not attract each other.

Quantized mass has already been detected in measurements of the International Prototype Kilogramme. This is a bar of 90% platinum and 10% iridium, kept in a vault on the outskirts of Paris, precisely made to provide a standard of mass. Weighings of the IPK are mysteriously imprecise; they vary by about one Planck mass or $2 \times 10^{-8} kg$. This indicates that mass is quantized.

The true *inertial* mass of the IPK is somewhere between multiples of the Planck mass. The most precise balance ever built could not measure more precisely than the Planck mass, because gravitational mass is quantized. Experimenters at this writing still haven't figured this out. They have promoted a new standard of mass based upon the Speed of Light.

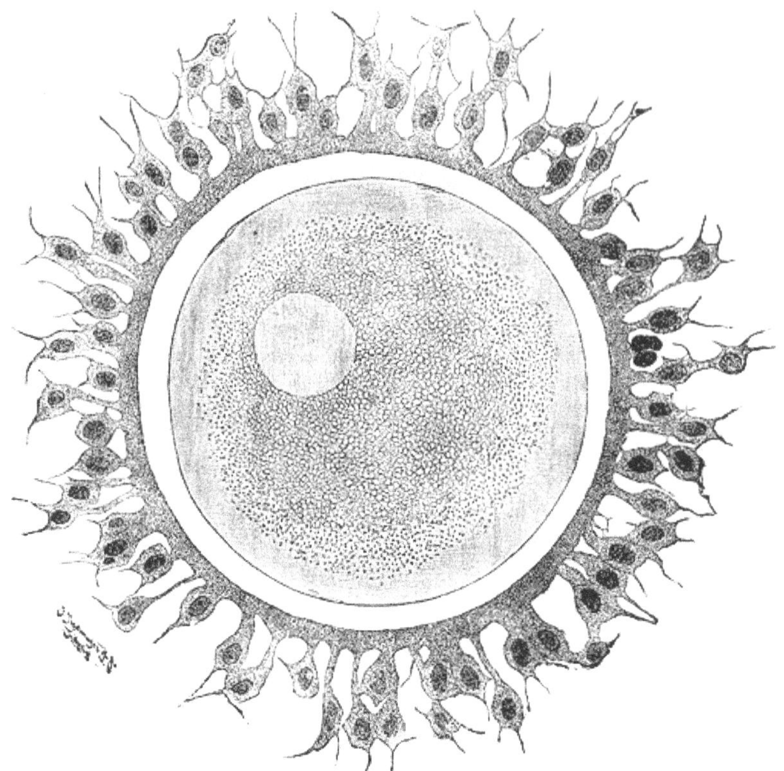

Most cosmologists' speculations have little bearing on our everyday life, but quantized gravity affects the size of cells in our body. As Robert Hooke saw in a microscope, all large forms of life are divided into cells. The tiny one-celled organisms seen in a microscope are unhindered by self-gravity. Why cells are this size has been a mystery, but all cells in our bodies are below the Planck mass.

The largest cell in the human body is the female egg cell, with a mass just under the Planck mass. When our egg cell is fertilized by a male sperm cell, the process of mitosis begins. During mitosis, genetic material migrates to opposite ends of the cell before dividing to create new cells.

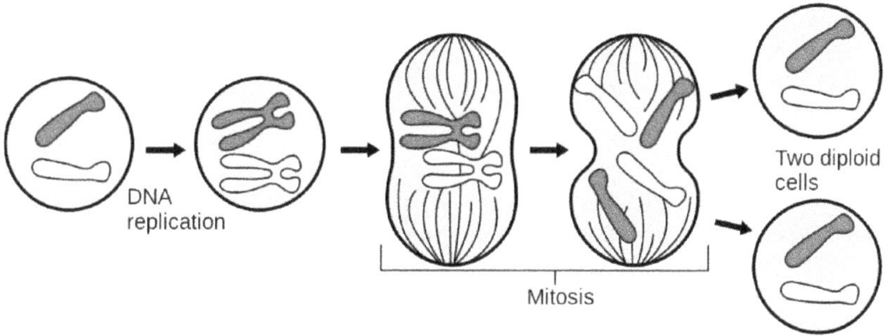

If cells were larger than the Planck mass, self-gravity would affect these internal workings.

A few life forms have cells that grow larger than the Planck mass. These cells have multiple nuclei to compensate for their size. The vast majority of cells, including those of the human body, are limited by the Planck mass. The size of our cells is a clue that gravitational mass is quantized at the Planck scale.

In the movie *Interstellar* Michael Caine, as NASA's Chief Scientist, spends a lifetime trying to link quantum mechanics with gravity but never succeeds. This is a prize that has escaped scientists from the time of Albert Einstein to today. It is seen as key to the next stage in understanding the Universe. Just as quantum mechanics has led to nuclear power, lasers and transistors, understanding a quantized gravity may someday lead to being able to control it.

Albert Einstein wrote, *"All physical theories, their mathematical expressions notwithstanding, ought to lend themselves to so simple a description that even a child could understand them."* This book begins when a child's eyes first open to light, after she has grown from a tiny cell less than a millionth of a meter across. Size of the entire Universe may be related to the size of that cell. The equations that describe our Universe can be understood by a child.

The Universe began and grew from a burst of light. The Speed of Light is a gateway to understanding mysteries of the Universe. Light brings discovery while "dark" energies spring from ignorance. As I write, humans may open their eyes to light.

LOUISE RIOFRIO
Hawaii

ABOUT THE AUTHOR

LOUISE RIOFRIO is educated in physics and astronomy, and worked as a Scientist at NASA Johnson Space Center in Houston. While studying the Moon, she performed experiments with Apollo lunar samples that had been untouched over 40 years. She is world-famous for predicting change in The Speed of Light, then using the Moon to measure change. She lives in the Hawaiian Islands and is invited to speak at scientific meetings worldwide.

Bibliography

Chapter 2
Ptolemy, *Geographia* and *Almagest*

Copernicus, Nicholas, *De Revolutionibus Orbium Coelestium,* 1543

Galilei, G., *The Starry Messenger,* 1610

Galilei, G., *Dialogue of the Two World Systems,* 1632

Chapter 3
Newton, Isaac, *Philosophia Naturalis Principia Mathematica,* 1686

Newton, Isaac, *Opticks,* 1704

Hooke, Robert, *Micrographia,* 1665

Thomson, W. and Tait, P.G. *Natural History*, Vol. 1, No. 403, 1874

Chapter 4
Einstein, Albert, "On a Heuristic Viewpoint Concerning the Production and Transformation of Light," *Annalen der Physik* Vol. 17, No. 6, 1905

Einstein, Albert, "On the Electrodynamics of Moving Bodies," *Annalen der Physik,* Vol. 17, No. 10, 1905

Zeeman, E.C., "Causality Implies the Lorentz Group," *Journal of Mathematical Physics,* Vol. 5, No. 490, 1964

Chapter 5
Einstein, Albert, *Relativity: The Special and General Theory,* Princeton University Press, 1961

Poe, Edgar Allen, *Eureka,* 1848

De Broglie, Louis, "Researches on the quantum theory," *Annalen der Physik,* Vol. 10, No. 3, 1925

Chapter 7
Riofrio, L., "$GM=tc^3$ Space/Time Explanation of Supernova Data," *Beyond Einstein*, 2004, http://www-conf.slac.stanford.edu/einstein/talks/aspauthor2004_3.pdf

Valentino, E., Melchiorri, A., and Silk, J., "Planck evidence for a closed universe and a possible crisis in cosmology," *Nature* Vol. 4, No. 2, p. 196-203, 2020

Sarkar, S. et al., "Evidence for anisotropy of cosmic acceleration," *Astronomy & Astrophysics,* Vol. 631, L13, 2019

Lee, Young-Wook et al., "Early-type Host Galaxies of Type Ia Supernovae II: Evidence for Luminosity Evolution in Supernova Cosmology," *Astrophysical Journal,* Vol. 889(1), No. 8, 2020

Chapter 8
Bills, B.G. and Ray, R.D., "Lunar Orbital Evolution: A Synthesis of Recent Results," *Geophysical Research Letters,* Vol. 26, 1999

Stephenson, F.R. and Morrison, L.V., "Longterm fluctuations in the Earth's rotation," *Philosophical Transactions of the Royal Society,* 1997

Poliakow, Eugene, "Numerical Modelling of the paleotidal evolution of the Earth-Moon system," *Proceedings of Intl Astronomical Union 197*, 2004

Riofrio, L., "Calculation of lunar orbit anomaly," *Planetary Science,* Vol. 1, No. 1, 2012, http://www.planetary-science.com/content/pdf/2191-2521-1-1.pdf

Chapter 9

Geller, Margaret, and Huchra, John, "Mapping the Universe," *Science* Vol. 246, No. 4932, 1989

Blanchard, Alain et al., "An alternative to the cosmological concordance model," *Astronomy and Astrophysics*, Vol. 412, No. 1, 2003

Chapter 10

Perelman, G., "Finite extinction time for the solutions to the Ricci flow on certain three-manifolds," 2003 http://arxiv.org/abs/math.DG/0307245

Dirac, Paul, "The Cosmological Constants," *Nature* Vol. 139, No. 3512, 1937

Eddington, Arthur, *Fundamental Theory,* Cambridge University Press, 1946

Chapter 11

Riofrio, L., "An Exceptionally Simple Experiment Testing Quantum Theory," *Journal of Cosmology,* Vol. 27, 2019, http://journalofcosmology.com/JOC27/contents/Quantized_Mass_Riofrio.pdf

Memories

Day 1

Memories

Day__

Memories

Day__

Memories

Day__

Memories

Day__

Memories

Day__

Memories

Day__

Memories

Day__

Memories

Day__

Memories

Day__

Memories

Day__

Day__

Photo

www.ingramcontent.com/pod-product-compliance
Ingram Content Group UK Ltd.
Pitfield, Milton Keynes, MK11 3LW, UK
UKHW061707190726
13853UKWH00008B/2443